ALGEBRA WORD PROBLEMS

Practice Workbook with Full Solutions

$$(x-4)(x+3)=0$$

Chris McMullen, Ph.D.

Algebra Word Problems Practice Workbook with Full Solutions
Improve Your Math Fluency
Chris McMullen, Ph.D.

www.monkeyphysicsblog.wordpress.com
www.improveyourmathfluency.com
www.chrismcmullen.wordpress.com

Zishka Publishing

ISBN: 978-1-941691-29-8

Textbooks > Math > Algebra
Study Guides > Workbooks> Math
Education > Math > Word Problems

CONTENTS

INTRODUCTION

This workbook is designed to help practice solving standard word problems. Every problem is fully solved using algebra: Simply turn the page to check the solution.

The first chapter offers some tips for solving word problems, the second chapter provides a quick refresher of essential algebra skills, and the third chapter includes several examples to help serve as a guide for how to solve algebra word problems.

A variety of problems are included, such as:

- age problems
- problems with integers
- relating the digits of a number
- fractions, decimals, and percentages
- average values
- ratios and proportions
- problems with money
- simple interest problems
- rate problems
- two moving objects
- mixture problems
- people working together
- problems with levers
- perimeter and area

May you (or your students) find this workbook useful and become more fluent with algebra word problems.

STRATEGIES AND TIPS

Read the Problem Carefully

First read the entire problem. Be sure to read every word. A single written word can make a big difference in the solution. It's a common mistake for students to focus so much on the numbers that they don't notice a very important word. As you read the problem, circle or underline what you believe will be key information:

- numbers like 12 years, $3.75, or 25%
- written numbers like two, one-third, or none
- key words that relate to mathematical operations like total, increased by, or tripled
- what you are solving for, like Anna's age or the number of apples in the cart

Identify the Given Information

The information given in the problem is used to solve for the desired unknown, so the first step is to gather the information that you know. You can do this by circling or underlining the numerical information in the problem, or you could make a table of this information.

- First identify all of the numbers like 3 bananas or 5 days.
- Beware that some numbers are stated using words, like writing "five" instead of 5, a "dozen" instead of 12, or "doubled" instead of 2 times.
- The number "zero" is often disguised. For example, if a problem states "there are no grapes left," this is equivalent to stating that there are zero grapes.

What Are You Solving for?

Read carefully to determine what the problem is asking you to find. Some problems ask a question like, "What is Julie's age?" or "How far did Pat walk?" Other problems state the question in a sentence like, "Determine the number of apples in the barrel." Some questions ask for more than one answer, like "How old are Liz and Tim?"

Indicate What Each Unknown Represents

In the beginning of the solution, it helps to write a phrase like the example below in order to remind you what each unknown represents. The unknown should usually represent what you are trying to solve for. That way, your solution will be complete once you solve for the unknown.

$$x = \text{the original number of cookies}$$

Multiple Unknowns

If there are two (or more) unknowns, try to let one variable represent the smallest unknown. For example, suppose that Melissa is three years older than Doug. In this case, Doug is younger, so you could let x represent Doug's age:

$$x = \text{Doug's age}$$
$$x + 3 = \text{Melissa's age}$$

If there are two (or more) unknowns, but it is difficult or inconvenient to express both unknowns in terms of a single variable (as we did above), it is possible to use two different variables. If you use multiple variables, you will need to write down more than one equation. If there are two variables, you will need two equations.

$$x = \text{bananas} \quad , \quad y = \text{oranges}$$
$$3x - 2y = 14$$
$$4x + 5y = 57$$

Relating the Unknowns to the Given Information

Write down an equation to help you solve for the variable. (If there are two different variables, you will need to write down two different equations.) The language in the problem helps you relate the variables to the given information. Translate the words into symbols by looking for words that relate to mathematical operations. Note that the examples in the following tables are designed to help with common expressions, but do not account for every possible way for the English language to describe each mathematical operation: You need to think about the wording of every problem.

Examples of Addition Language	Addition Examples with Unknowns	
sum, total of, increased by, gained, greater than, more than, raised to, combined, together, in all	8 more than a number	$x + 8$
	Katie's age 12 years from today	$x + 12$

Examples of Subtraction Language	Subtraction Examples with Unknowns	
difference, minus, decreased by, lost, less than, smaller than, fewer, left over, taken away, between, after	5 less than a number	$x - 5$
	Jack's age one year ago	$x - 1$

Examples of Multiplication Language	Multiplication Examples with Unknowns	
multiplied by, product, times, of, increased by a factor of, decreased by a factor of, twice, double, triple, each	7 times a number	$7x$
	twice as much	$2x$

Examples of Division/Fraction Language	Division/Fraction Examples with Unknowns	
divided by, out of, per, split, equal pieces, average, fraction, ratio, quotient, percentage, half, third, fourth	half as much	$\frac{x}{2}$
	25% of the students	$0.25x$ or $\frac{x}{4}$ (since $0.25 = \frac{1}{4}$)

Examples of Root/Power Language	Root/Power Examples with Unknowns	
squared, cubed, raised to the power of, squareroot, cube root, root	a number cubed	x^3
	the squareroot of a number	$\sqrt{x}$

Examples of Equal Sign Language	Equal Sign Example with Unknowns	
is, was, makes, will be, equals	the sum of two numbers is 15	$x + y = 15$

Beware of Possible Extraneous Information

Occasionally, a problem includes extraneous information that isn't needed to solve a problem. Although most problems give you only the information that is needed, it is a good habit to ask, "Which information is needed to solve the problem?" Remember that a rare problem may include numbers that aren't relevant to the solution.

Be Confident and Determined

Successful students know that a solution exists. They are determined to figure it out.

Working with Integers

The following features are common in word problems:

- Represent two consecutive integers with x and $(x + 1)$. A third consecutive integer would equal $(x + 2)$, and so on.
- Represent two consecutive even or odd integers with x and $(x + 2)$. (The two numbers will have a difference of 2 whether they are both odd or both even.) If there is a third consecutive even or odd integer, that equals $(x + 4)$.
- To solve for the digits of a two-digit number, multiply the tens digit by 10 and the units digit by 1. For example, if the problem states that the units digit is 5 times the tens digit, let the tens digit equal x, the units digit equals $5x$, and the number equals $10(x) + 1(5x) = 10x + 5x = 15x$. Suppose that you solve the problem and obtain $x = 1$. In this example, the tens digit is 1, and the units digit is 5. The number is $10(1) + 1(5) = 15$.
- To solve for the digits of a three-digit number, multiply the hundreds digit by 100, the tens digit by 10, and the units digit by 1. For example, if the problem states that the tens digit is twice the units digit and that the hundreds digit is triple the tens digit, let the units digit equal x, the tens digit equals $2x$, the hundreds digit equals $3(2x) = 6x$, and the number equals $100(6x) + 10(2x) + 1(x) = 600x + 20x + x = 621x$. Suppose that you solve the problem and obtain $x = 1$. In this example, the units digit is 1, the tens digit is 2, and the hundreds digit is 6. The number is $100(6) + 10(2) + 1(1) = 621$.
- To reverse the digits of a two-digit number, swap the place of the tens and units digit. For example, if a problem states that the units digit is x and the tens digit is $x + 2$, the number is $10(x + 2) + 1(x) = 10x + 20 + x = 11x + 20$ and the reversed number is $10(x) + 1(x + 2) = 10x + x + 2 = 11x + 2$. Suppose that you solve the problem and obtain $x = 5$. In this example, the units digit is 5, the tens digit is 7, the number is $10(7) + 1(5) = 75$ and the reversed number is $10(5) + 1(7) = 57$. Observe that 75 and 57 indeed have their digits reversed.

Sum, Product, Difference, and Ratio

If you know the sum, product, difference, or ratio of two numbers, but aren't told what either number equals, let the following examples serve as a guide:

- If the sum of two numbers equals 42 (for example), let one number be x and the other number will be $(42 - x)$.
- If the product of two numbers equals 36 (for example), let one number be x and the other number will be $\frac{36}{x}$.
- If the difference between two numbers is 5 (for example), let one number be x and the larger number will be $(x + 5)$.
- If the ratio of two numbers is 3 (for example), let one number be x and the larger number will be $3x$.

Fractions, Decimals, and Percentages

Following are some tips for dealing with fractions, decimals, and percentages:

- Divide by 100 to convert a percent into a decimal.

$$40\% = \frac{40}{100} = 0.4$$

- When there are decimals in an equation, multiply the entire equation by the power of 10 needed in order to remove all of the decimals.

$$0.24x + x = 6 \quad \to \quad \text{multiply by } 100 \quad \to \quad 24x + 100x = 600$$

- When there are fractions in an equation, multiply the entire equation by the lowest common denominator.

$$\frac{x}{2} - \frac{1}{x} = \frac{1}{3} \quad \to \quad \text{multiply by } 6x \quad \to \quad 3x^2 - 6 = 2x$$

- The phrases "increased by" or "decreased by" are compared to 100% (or 1).
  - If x increases by 20%, this means $1.2x$ (since $120\% = 1.2$).
  - If x decreases by 1/4, this means $\frac{3x}{4}$ or $0.75x$ (since $1 - \frac{1}{4} = \frac{3}{4} = 0.75$).

Ratios and Proportions

A ratio expresses a fixed relationship in the form of a fraction. For example, if there are 300 girls and 200 boys in a particular school, the ratio of girls to boys attending that school is 3 to 2. We could express this ratio with a colon (3:2), as a fraction $\left(\frac{3}{2}\right)$, as a decimal (since $\frac{3}{2} = 1.5$), or as a percent (150%).

When a problem gives you the ratio, but not the quantity of each, if you let x be the quantity represented by the denominator, multiply the ratio by x to get the quantity represented by the numerator. For example, if the ratio of white cars to black cars is 5:4, if you let x represent the number of black cars (since black corresponds to the denominator), the number of white cars will be $1.25x$ (since $5{:}4 = \frac{5}{4} = 1.25$).

A proportion expresses an equality between two ratios. For example, if the ratio of apples to oranges is 4:3 and there are 64 apples, if we let x represent the number of oranges, we can use the following proportion to solve for x. Check for consistency when setting up a proportion: On both sides of the following equation, apples are on top and oranges are on the bottom.

$$\frac{4}{3} = \frac{64}{x}$$

Cross multiply in order to remove the variable from the denominator.

$$4x = 3(64) = 192$$
$$x = \frac{192}{4} = 48$$

In this example, there are $x = 48$ oranges and $\frac{4}{3}x = \frac{4}{3}(48) = 64$ apples. As a check, note that $\frac{64}{48} = \frac{4}{3}$, such that the ratio of apples to oranges is indeed 4:3.

Money and Interest

For problems with money expressed in decimals, like \$3.25, after you write down the equation, if you multiply both sides of the equation by 100, it will remove all of the decimals from the problem. See the example below.

$$2.25x - 42.97 = 12x \quad \rightarrow \quad 225x - 4297 = 1200x$$

For problems that involve US coins, it is often convenient to express the money in terms of cents. For example, $5x + 10y$ is the amount of cents contained in x nickels and y dimes, since each nickel is worth 5 cents and each dime is worth 10 cents.

penny	**nickel**	**dime**	**quarter**
1 cent	5 cents	10 cents	25 cents

For problems that involve simple interest calculations, note that the interest (I) is equal to the principal (P) times the interest rate (r) in decimal form. In the formula below, note that P is multiplying r.

$$I = Pr$$

For example, suppose that a student invests \$500 in a savings account that earns interest at a rate of 3%. The principal is the original amount invested: $P = \$500$. Divide the interest rate by 100% to convert it into a decimal: $r = \frac{3\%}{100\%} = 0.03$. Use the formula above to determine the interest earned.

$$I = Pr = (\$500)(0.03) = \$15$$

If the account earns 3% interest per year, after one year, the new balance will be \$515 (add the original principal to the interest to determine this). To determine the interest earned after two years, for the second year use \$515 as the new principal.

Average Values

For simple averages (defined as the arithmetic mean), apply the following formula. Add up all of the values ($V_1, V_2, \ldots, V_N$) and divide by the number of values (N).

$$V_{avg} = \frac{V_1 + V_2 + \cdots + V_N}{N}$$

For example, consider the values 18, 22, and 23. In this example, $N = 3$ since there are 3 different values. According to the formula, the average value of these numbers is:

$$V_{avg} = \frac{18 + 22 + 23}{3} = \frac{63}{3} = 21$$

Check that your average value lies somewhere in between the smallest and greatest values. In this example, the smallest value is 18, the greatest value is 23, and the average value of 21 lies in between them.

Constant Rates

When the rate is constant, the rate (r) equals the distance (d) traveled divided by the time (t) taken. The units must be consistent. For example, if the rate is given in kilometers per hour, you want the distance to be in kilometers (not meters or miles) and the time to be in hours (not minutes or seconds), but if the rate is given in feet per second, you want the distance to be in feet and the time to be in seconds. If the given units aren't consistent, you will need to perform a unit conversion before you solve the problem. It may help to recall that there are 60 seconds in one minute, 60 minutes in one hour, 3 feet in one yard, 1760 yards in one mile, and 1000 meters in one kilometer.

$$r = \frac{d}{t}$$

Two Moving Objects

If there are two objects moving with constant rates, first organize the information for each object into a table. Some of this information will be numerical values stated in the problem. The rest will be expressed in terms of a variable (such as x). Define precisely in words what each variable (like x) represents, and express any unknown quantities in terms of this variable.

For example, suppose that a boy and a girl are initially 25 m apart and begin walking towards one another at the same moment. The boy walks with a constant speed of 2 m/s while the girl walks with a constant speed of 3 m/s. They continue walking until they meet. We organized this information in the table below. The boy and girl travel for the same amount of time in this example (since they start and finish at the same time). The speeds are the rates. Since rate equals distance divided by time ($r = \frac{d}{t}$), it follows that distance equals rate times time ($d = rt$). Therefore, the boy travels $2t$ and the girl travels $3t$. In this example, $2t + 3t = 25$ is the total distance traveled.

	distance traveled (in meters)	time traveled (in seconds)	rate (in m/s)
boy	$2t$	t	2
girl	$3t$	t	3

Don't memorize how the table looks, since the table will look somewhat different for different problems. Instead, try to learn how to read a problem and reason out how to enter the information in the table. Study the table above and study Examples 15-17 in Chapter 3.

Do the two objects travel the same distance, or do they travel for the same amount of time? Usually, one of these is the same for both objects, but not both.

Do two objects travel the same distance?

- If both objects begin in the same place and also finish in the same place, and if both objects also travel along the same path, then the two objects travel the same distance.
- If the objects start in different positions, finish in different places, or travel along different paths, the distance traveled may be different for each object.
- If the objects travel for the same time, the distances are usually different.

Does one object have a head start?

- If one object starts before the other object, the time traveled will be shorter for the object that starts last.
- For example, suppose that a son starts running 2 seconds before his father starts running. If you let t represent the time that the father runs, then $t + 2$ will be the time that the son runs (since the son spends more time running).

Do two objects travel for the same amount of time?

- If both objects begin moving at the same time (or if both objects are already moving when the problem begins) and also finish (or meet up) at the same time, then the total time is the same for both objects.
- If one object starts before the other object, then the time traveled is different for each object.
- If the objects travel the same distance, the times are usually different.

It often helps to draw a diagram. Draw and label the following:

- the initial and final points for each object
- the path that each object takes
- given information, such as distance, time, or rate for each object

Mixtures

The amount (P) of pure substance contained in a given amount (M) of a mixture is given by the following formula, where the concentration (c) is the decimal form of a percentage. Divide by 100% to convert a percentage to a decimal.

$$P = cM$$

For example, suppose that 20 liters of a solution is 25% ethanol (by volume). In this example, $c = 0.25$ (in decimal form), $M = 20$ liters, and the amount of pure ethanol is $P = cM = (0.25)(20) = 5$ liters. (Note that in chemistry, concentration may be expressed as volume/volume, mass/volume, moles/volume, mass/mass, etc. That is, the amounts P and M do not always have units of volume. Don't worry: You won't need to know any chemistry to solve the word problems in this book. We will focus on the algebra, not on the science.)

Combining Two Mixtures Together

The following formulas apply when two different solutions are mixed together. The first formula states that the total amounts of the two solutions add up to the total amount of the combined solution. The second formula states that the total amount of pure substance in the two solutions add up to the total amount of pure substance in the combined solution. (Beware that the percentages do **not** add up like this.)

$$M_1 + M_2 = M_3$$
$$P_1 + P_2 = P_3$$

The formula $P = cM$ applies to each solution (1, 2, or 3). For example, when it is applied to the 3rd solution (the combined solution), we get $P_3 = c_3 M_3$. Since there are 3 different solutions, the subscripts 1, 2, and 3 help you distinguish between them. The example on the following page shows you how to apply these formulas to a problem involving a mixture of solutions.

If two solutions are mixed together, first organize the information for each solution into a table. Some of this information will be numerical values stated in the problem. The rest will be expressed in terms of a variable (such as x). Define precisely in words what each variable (like x) represents, and express any unknown quantities in terms of this variable.

For example, suppose that 4 liters of 15% sulfuric acid is mixed with 2 liters of 10% sulfuric acid. We organized this information in the table below. We let x represent c_3 (the percent of sulfuric acid in the mixed solution, but in decimal form). Note that $M_1 + M_2 = M_3$ becomes $4 + 2 = 6$ and $P_1 + P_2 = P_3$ becomes $0.6 + 0.2 = 6x$. The last row was made by applying the formula $P = cM$ to each solution: Multiply the second row by the third row to get the bottom row.

	15% sulfuric acid	10% sulfuric acid	the two solutions mixed together
c (decimal)	0.15	0.10	x
M (liters)	4	2	$4 + 2 = 6$
P (pure sulfuric acid)	$(0.15)(4) = 0.6$	$(0.1)(2) = 0.2$	$6x$

We will explore how to make and apply the above table in Examples 19-21 of Chapter 3. The tables will vary somewhat from one problem to another, though the reasoning for how to complete and apply the table is essentially the same. Strive to understand the logic behind making the table. Don't try to memorize each table.

If a problem adds water to a solution to dilute the solution, note that c will be zero for water. If a problem adds a pure substance (like pure sulfuric acid) to a mixture, note that c will be 1 (for 100%) for a pure substance.

People Working Together

Suppose that two (or more) people are doing the same job, but that one person does the work faster than the other. First organize the information for each worker into a table. Some of this information will be numerical values stated in the problem. The rest will be expressed in terms of a variable (such as x). Define precisely in words what each variable (like x) represents, and express any unknown quantities in terms of this variable.

The key to solving such problems is to work with fractions. Begin with the time it would take each person to complete the job individually. Next take the reciprocal of these times to determine the fraction of the work that each worker could complete in a single time period. For example, if Jenny can complete a job in 5 hours, then in a single hour she could complete $\frac{1}{5}$ of the job. (The reciprocal of a number is found by dividing 1 by that number.)

Be sure to use the same units of time, like hours or minutes, for each worker. Don't mix and match the units. Recall that there are 60 seconds in one minute, 60 minutes in one hour, 24 hours in one day, 7 days in one week, 52 weeks in one year, and 12 months in one year.

For example, suppose that Fred can eat a whole pie in 12 minutes while Ryan can eat a whole pie in 9 minutes. We organized this information in the table below. We let x represent the time it would take for Fred and Ryan to finish one pie together.

	Fred	**Ryan**	**together**
time to eat (in minutes)	12	9	x
fraction eaten per minute	$\frac{1}{12}$	$\frac{1}{9}$	$\frac{1}{x}$

The reason for working with fractions is that the fractions add up: In the previous example, $\frac{1}{12} + \frac{1}{9} = \frac{1}{x}$. To perform the algebra, first make a common denominator. The lowest common denominator of $\frac{1}{12}$ and $\frac{1}{9}$ is $\frac{1}{36}$. Note that $\frac{1}{12} = \frac{1}{12}\frac{3}{3} = \frac{3}{36}$ and $\frac{1}{9} = \frac{1}{9}\frac{4}{4} = \frac{4}{36}$, such that $\frac{1}{12} + \frac{1}{9} = \frac{3}{36} + \frac{4}{36} = \frac{7}{36} = \frac{1}{x}$. Now take the reciprocal of both sides to get $\frac{36}{7} = \frac{x}{1} = x$ (since $\frac{x}{1} = x$). Alternatively, after we found the common denominator we could have cross multiplied to solve for x. We would still get $x = \frac{36}{7}$.

Check that your answer makes sense: If two people work on a project together, they will finish the project sooner than if either person worked on the project alone.

Torque and Levers

A torque is exerted when a force is applied that would tend to cause rotation. The formula for torque is force times lever arm. The equation for torque explains why it would be difficult to open a door if the handle were located near the hinges: In that case, the lever arm would be smaller, resulting in less torque.

The lever illustrated below consists of a long bar. The lever is resting on a fulcrum: This is the balancing point, illustrated as a small triangle in the figure. Two boxes are placed on the bar. The boxes have different weights (w_1 and w_2) and are different distances from the fulcrum (r_1 and r_2).

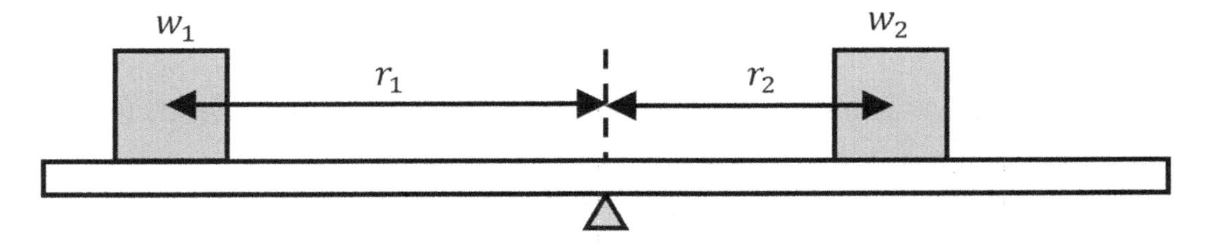

For a lever in static equilibrium, the sum of the clockwise torques equals the sum of the counterclockwise torques. In the figure above, the right box exerts a clockwise torque and the left box exerts a counterclockwise torque. Note that each torque is equal to weight times lever arm (where lever arm is the distance from the fulcrum).

For the example shown on the previous page:

$$w_1 r_1 = w_2 r_2$$

If there are only two weights in a lever problem, the heavier object will be closer to the fulcrum. That is, a larger value of w has a corresponding smaller value of r.

If there are additional weights on either side of the lever, add additional torques to the corresponding side of the equation. For example, with two weights on the left and one on the right, the equation would be:

$$w_1 r_1 + w_2 r_2 = w_3 r_3$$

Strictly speaking, weight (w) is a force measured in Newtons or pounds (where lb. is the symbol for the pound), whereas mass is measured in grams. However, for lever problems it is okay to work with mass instead of weight. Why? Weight equals mass times gravitational acceleration, where gravitational acceleration is 9.81 m/s^2 near earth's surface. Every term in the lever equation includes this same numerical value, so the equation will still be true if we divide the entire equation by 9.81 m/s^2. What does this mean? It means that it doesn't matter whether the problem gives weights in Newtons or pounds or whether it gives masses in grams or kilograms: Either way, you can still use the same lever equation (and there is no reason to use the number 9.81 m/s^2 in the calculations, since it would just cancel out anyway).

Examples 25-27 in Chapter 3 involve lever problems.

Note: In this book, we will neglect the weight of the lever (the bar or rod) unless the problem specifically states otherwise.

Perimeter and Area

The following formulas may be helpful for problems involving common geometric shapes. Note that the height (h) of a triangle is perpendicular to the base (b), the diameter (D) of a circle passes through its center, the radius (r) of a circle extends from its center, and the circumference (C) of a circle is the distance around its edge. The constant π equals the ratio of the circumference of a circle to its diameter, and is approximately equal to 3.14159.

Triangle (sides a, b, c; height h; base b)	area	$A = \frac{1}{2}bh$
	perimeter	$P = a + b + c$
Rectangle 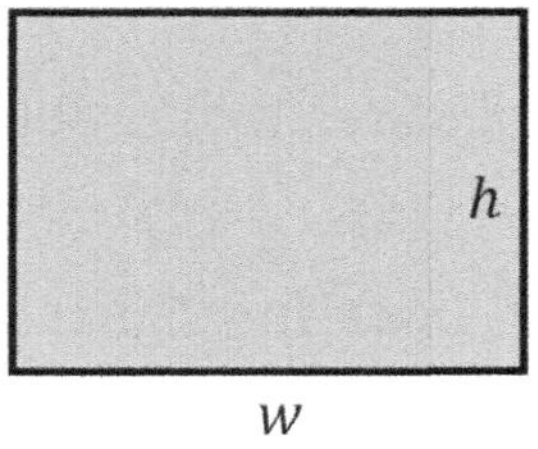 (width w, height h)	area	$A = wh$
	perimeter	$P = 2w + 2h$
Circle 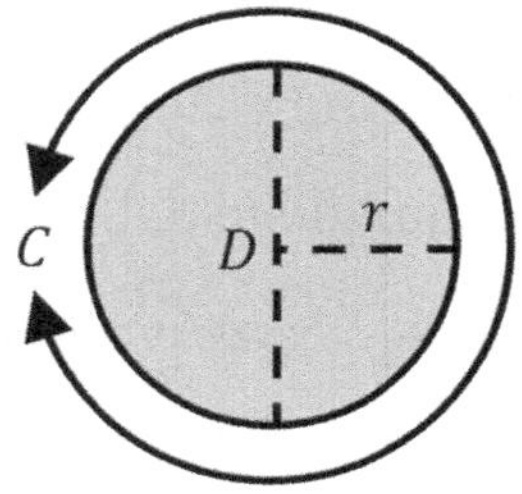 (circumference C, diameter D, radius r)	diameter	$D = 2r$
	area	$A = \pi r^2$
	circumference	$C = 2\pi r$

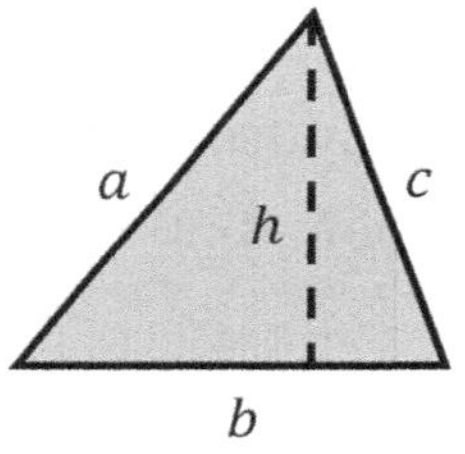

Solve for the Unknown

Once you translate the word problem into one or more equations, all that remains is to apply your algebra skills in order to solve for the unknown(s). You should know from an algebra class how to isolate an unknown, how to distribute, how to factor, how to apply the quadratic equation, and how to solve a system of equations. The chapter that follows provides a quick review of essential algebra skills.

Strategy for Solving Word Problems

To solve a word problem, follow these steps:

- Read the problem carefully. Circle or underline key information.
- Identify the given information (including any numbers that may be written in the form of words).
- What are you solving for?
- Indicate clearly what each unknown represents. Write this out in words, like "x = Mark's age" or "t = time."
- If there are two (or more) people, objects, or mixtures in a problem, make a table to help organize the information.
- Relate the unknowns to the given information. Look for signal words (like "difference" for subtraction or "makes" for equals).
- If necessary, review the relevant sections from this chapter. For example, if a problem involves interest, review the section called Money and Interest.
- Some problems involve specific formulas. For example, for a moving object apply the rate equation, or to find the area of a triangle use $A = \frac{1}{2}bh$.
- Study the relevant examples from Chapter 3. These serve as a helpful guide.
- Once you have one equation with a single unknown (or 2 equations with 2 unknowns), apply algebra to solve for the unknown (see Chapter 2).
- Think about your answer. Check that your answer makes sense.

ALGEBRA REFRESHER

Combine Like Terms

Terms with the same variable raised to the same power are like terms. For example, $3x$ and $2x$ are like terms, x^2 and $5x^2$ are like terms, and $2y$ and $-y$ are like terms.

$$x^2 + 3x + 2y + 2x + 5x^2 - y = 6x^2 + 5x + y$$

Numerical constants like 5, 18, and -6 are like terms.

$$x^2 + 5 + 18 + 3x - 6 = x^2 + 23 + 3x - 6 = x^2 + 3x + 17$$

Isolate the Unknown

In some algebraic equations, it is possible to isolate the unknown by combining like terms with the variables on one side and the constants on the other side. Apply the same operation to both sides of the equation. This is shown in the example below.

$$8x - 60 = 5x - 12$$

Add 60 to both sides. Subtract $5x$ from both sides.

$$8x - 5x = 60 - 12$$

$$3x = 48$$

Divide both sides by 3.

$$x = \frac{48}{3} = 16$$

Distributing

When two (or more) terms in parentheses multiply (or divide by) an algebraic term, the operation (multiplication or division) distributes to each term, like the examples below. This is the distributive property of multiplication.

$$4x(3x^2 - 5) = 4x(3x^2) + 4x(-5) = 12x^3 - 20x$$

$$-(8x - 3) = -8x - (-3) = -8x + 3$$

$$\frac{6x^2 + 8x}{2x} = \frac{6x^2}{2x} + \frac{8x}{2x} = 3x + 4$$

The f.o.i.l. Method

The abbreviation f.o.i.l. stands for first, outside, inside, last. This abbreviation helps to remember how to multiply an expression like this:

$$(w + x)(y + z) = wy + wz + xy + xz$$

The following identity is often useful.

$$(x + y)(x - y) = x^2 - xy + yx - y^2 = x^2 - y^2$$

That little minus makes a big difference. Compare with the following.

$$(x + y)(x + y) = x^2 + xy + yx + y^2 = x^2 + 2xy + y^2$$

Factoring

When you apply the distributive property backwards, it is called factoring.

$$6x^6 + 8x^9 = 2x^3(3x^3 + 4x^6)$$

$$\frac{7x}{2} - \frac{3y}{2} = \frac{7x - 3y}{2}$$

Powers

Recall the following rules concerning powers (or exponents).

$x^0 = 1 \quad (\text{if } x \neq 0)$	$x^1 = x$	$x^{-1} = \frac{1}{x}$	$\frac{1}{x^{-1}} = x$
$x^m x^n = x^{m+n}$	$(ax)^m = a^m x^m$	$(x^m)^n = x^{mn}$	$\frac{1}{x^{-m}} = x^m$
$\frac{x^m}{x^n} = x^{m-n}$	$\left(\frac{1}{x^m}\right)^n = \frac{1}{x^{mn}}$	$\left(\frac{a}{x}\right)^m = \frac{a^m}{x^m}$	$x^{-m} = \frac{1}{x^m}$

Fractional Powers

Recall the following rules concerning fractional powers (or exponents).

$x^{1/2} = \sqrt{x}$	$x^{-1/2} = \frac{1}{\sqrt{x}}$	$(ax)^{1/2} = \sqrt{ax}$	$x^{m/n} = \left(\sqrt[n]{x}\right)^m$

Squareroots

Recall the following rules concerning squareroots.

$\sqrt{x}\sqrt{x} = x$	$\frac{1}{\sqrt{x}} = \frac{\sqrt{x}}{x}$	$\frac{x}{\sqrt{x}} = \sqrt{x}$	$\sqrt{ax} = \sqrt{a}\sqrt{x}$

Negative Numbers

Recall the following rules concerning negative numbers.

$x - (-y) = x + y$	$x(-y) = -xy$	$(-x)(-y) = xy$	$\frac{-x}{-y} = \frac{x}{y}$

The Quadratic Formula

An equation of the form $ax^2 + bx + c = 0$ can be solved via the quadratic formula:

$$x = \frac{-b \pm \sqrt{b^2 - 4ac}}{2a}$$

Factoring the Quadratic

The example below shows how a quadratic equation can sometimes be factored.

$$3x^2 - 10x - 8 = 0$$

$$(3x + 2)(x - 4) = 0$$

$$3x + 2 = 0 \quad \text{or} \quad x - 4 = 0$$

$$x = -\frac{2}{3} \quad \text{or} \quad x = 4$$

Common Denominator

The way to add or subtract fractions is to make a common denominator. This is true even when there is a variable involved.

$$\frac{1}{2} - \frac{1}{3} = \frac{1}{2}\frac{3}{3} - \frac{1}{3}\frac{2}{2} = \frac{3}{6} - \frac{2}{6} = \frac{3-2}{6} = \frac{1}{6}$$

$$\frac{x}{3} + \frac{x}{4} = \frac{x}{3}\frac{4}{4} + \frac{x}{4}\frac{3}{3} = \frac{4x + 3x}{12} = \frac{7x}{12}$$

$$\frac{1}{x} + \frac{3}{x^2} = \frac{1}{x}\frac{x}{x} + \frac{3}{x^2} = \frac{x}{x^2} + \frac{3}{x^2} = \frac{x+3}{x^2}$$

$$\sqrt{x} + \frac{1}{\sqrt{x}} = \sqrt{x}\frac{\sqrt{x}}{\sqrt{x}} + \frac{1}{\sqrt{x}} = \frac{\sqrt{x}\sqrt{x} + 1}{\sqrt{x}} = \frac{x+1}{\sqrt{x}}$$

Dividing by Fractions

To divide by a fraction, multiply by its reciprocal. The reciprocal of $\frac{x}{y}$ is $\frac{y}{x}$. (Note that the reciprocal of x is $\frac{1}{x}$ because $\frac{x}{1} = x$.)

$$\frac{x}{y/z} = x\left(\frac{z}{y}\right) = \frac{xz}{y}$$

$$\frac{w/x}{y/z} = \frac{w}{x}\left(\frac{z}{y}\right) = \frac{wz}{xy}$$

$$\frac{1}{1/x} = x$$

$$\frac{x/y}{z} = \frac{x/y}{z/1} = \frac{x}{y}\left(\frac{1}{z}\right) = \frac{x}{yz}$$

Cross Multiply

The following example illustrates how to cross multiply.

$$\frac{w}{x} = \frac{y}{z}$$

$$wz = xy$$

Variable in a Denominator

If there are 3 (or more) terms, solve for the unknown term and take a reciprocal.

$$\frac{1}{x} + \frac{1}{12} = \frac{1}{3} \quad \rightarrow \quad \frac{1}{x} = \frac{1}{3} - \frac{1}{12} = \frac{1}{3}\frac{4}{4} - \frac{1}{12} = \frac{4-1}{12} = \frac{3}{12} = \frac{1}{4}$$

Now take the reciprocal of both sides of $\frac{1}{x} = \frac{1}{4}$.

$$x = 4$$

Systems of Equations

If there are 2 equations with 2 unknowns (or 3 equations with 3 unknowns), one way to solve the system is by substitution, like the example below.

$$2x + y = 12 \quad , \quad 4x - 3y = 14$$

$$2x + y = 12 \quad \to \quad y = 12 - 2x$$

$$4x - 3y = 14 \quad \to \quad 4x - 3(12 - 2x) = 14 \quad \to \quad 4x - 36 + 6x = 14$$

$$\to \quad 10x = 50 \quad \to \quad x = \frac{50}{10} = \boxed{5}$$

$$y = 12 - 2x = 12 - 2(5) = 12 - 10 = \boxed{2}$$

A system of equations can also be solved simultaneously, like the example below.

$$5x - 4y = 19 \quad \text{multiply by 3} \quad \to \quad 15x - 12y = 57$$

$$3x + 6y = 45 \quad \text{multiply by 2} \quad \to \quad 6x + 12y = 90$$

Add the equations: $15x + 6x = 57 + 90 \quad \to \quad 21x = 147 \quad \to \quad x = \frac{147}{21} = \boxed{7}$

$$3(7) + 6y = 45 \quad \to \quad 21 + 6y = 45 \quad \to \quad 6y = 24 \quad \to \quad y = \frac{24}{6} = \boxed{4}$$

Factor Perfect Squares

When an answer involves a squareroot, if there are any perfect squares inside of the squareroot, it is customary to factor them out like the example below.

$$x = \sqrt{48} = \sqrt{(16)(3)} = \sqrt{16}\sqrt{3} = 4\sqrt{3}$$

Rationalize the Denominator

When there is a squareroot in the denominator of an answer, it is conventional to rationalize the denominator like the example below.

$$x = \frac{1}{\sqrt{3}} = \frac{1}{\sqrt{3}}\frac{\sqrt{3}}{\sqrt{3}} = \frac{\sqrt{3}}{3}$$

EXAMPLES

These examples are designed to help serve as a guide for how to apply algebra to solve a variety of standard word problems. Each topic included in this workbook has examples in this chapter.

Note that there is more to solving problems than simply finding the right example and changing the numbers. A good problem-solver must learn to think through the problems and learn not to rely too much on the examples. It isn't practical to include an example of every possible type of problem because the book would be thousands of pages long.

Good problem-solvers don't try to copy, mimic, or memorize examples. Rather, they read through the examples and try to understand the logic, reasoning, and ideas that are involved in the solution. Once they understand how to solve a few representative examples, good problem-solvers can adapt the logical reasoning to other problems that are not quite the same as the examples.

Anybody can become a good problem-solver. The "trick" is to think your way through the examples and solutions. Try to understand the "why" and the "how." Try to follow the logic, reasoning, and underlying ideas. Study the solution to each problem. Learn from any mistakes that you made. Practice solving a variety of problems. Approach each problem confidently, but with open-mindedness.

Beware that a problem may apply something that you know in a new way. The more you understand a concept, the better your chances of recognizing it out of its usual context. Being able to solve problems that are applied in different ways is a valuable skill, which is why exam-makers frequently employ this tactic.

EXAMPLE #1

Two numbers have a sum of 25 and a product of 136. What are the numbers?

SOLUTION

Define the variable clearly:

$$x = \text{the first number}$$

The sum equals 25. This means that the two numbers add up to 25. Since the first number is x, it follows that:

$$25 - x = \text{the second number}$$

The product equals 136. When the numbers are multiplied together, they make 136.

$$x(25 - x) = 136$$

Distribute the x.

$$25x - x^2 = 136$$

This is a quadratic equation. Reorder the terms.

$$-x^2 + 25x - 136 = 0$$

Compare this to the standard form, $ax^2 + bx + c = 0$, to see that $a = -1$, $b = 25$, and $c = -136$. Plug these values into the quadratic formula.

$$x = \frac{-b \pm \sqrt{b^2 - 4ac}}{2a} = \frac{-25 \pm \sqrt{25^2 - 4(-1)(-136)}}{2(-1)} = \frac{-25 \pm \sqrt{625 - 544}}{-2}$$

$$x = \frac{-25 \pm \sqrt{81}}{-2} = \frac{-25 \pm 9}{-2}$$

There are two possible answers corresponding to the plus and minus signs.

$$x = \frac{-25 + 9}{-2} = \frac{-16}{-2} = 8 \quad \text{or} \quad x = \frac{-25 - 9}{-2} = \frac{-34}{-2} = 17$$

If the first number is $x = 8$, then the second number is $25 - 8 = 17$. Thus, we see that the two numbers are $\boxed{8}$ and $\boxed{17}$. Check the answers: 8 and 17 have a sum of $8 + 17 = 25$ and have a product of $(8)(17) = 136$.

EXAMPLE #2

One number is five times another number. The numbers have a difference of 36. What are the numbers?

SOLUTION

Define the variable clearly:

$$x = \text{the smaller number}$$

The difference equals 36. This means that one number is 36 more than the other number. Since the smaller number is x, it follows that:

$$x + 36 = \text{the larger number}$$

The larger number is five times the smaller number. Multiply the smaller number (x) by 5 and set this product equal to the larger number ($x + 36$).

$$5x = x + 36$$

Isolate the unknown. Combine like terms: Subtract x from both sides.

$$5x - x = 36$$

$$4x = 36$$

$$x = \frac{36}{4} = 9$$

The smaller number is $x = 9$ and the larger number is $x + 36 = 9 + 36 = 45$. The two numbers are $\boxed{9}$ and $\boxed{45}$. Check the answers: The difference between 45 and 9 is 36, and 9 times 5 makes 45.

EXAMPLE #3

Three consecutive odd numbers add up to 99. What are the numbers?

SOLUTION

Define the variable clearly:

$$x = \text{the smallest number}$$

Consecutive odd numbers have a difference of 2. Since the smaller number is x, it follows that:

$$x + 2 = \text{the middle number}$$

$$x + 4 = \text{the largest number}$$

The three numbers add up to 99. Add the three numbers together.

$$x + (x + 2) + (x + 4) = 99$$

Isolate the unknown. Combine like terms.

$$x + x + 2 + x + 4 = 99$$

$$3x + 6 = 99$$

$$3x = 99 - 6$$

$$3x = 93$$

$$x = \frac{93}{3} = 31$$

The smallest number is $x = 31$, the middle number is $x + 2 = 31 + 2 = 33$, and the largest number is $x + 4 = 31 + 4 = 35$. The three numbers are $\boxed{31}$, $\boxed{33}$, and $\boxed{35}$. Check the answers: 31, 33, and 35 are consecutive odd numbers and their sum is $31 + 33 + 35 = 99$.

EXAMPLE #4

William is three times as old as Susan. Six years from now, William will be twice as old as Susan. What are their ages now?

SOLUTION

Define the variable clearly. Susan is younger than William.

$$x = \text{Susan's age now}$$

William is three times as old as Susan. This involves multiplication.

$$3x = \text{William's age now}$$

The second sentence refers to their ages six years from now. Add 6 to their current ages to find their ages 6 years from now.

$$x + 6 = \text{Susan's age 6 years from now}$$

$$3x + 6 = \text{William's age 6 years from now}$$

Six years from now, William will be twice as old as Susan. Multiply Susan's age by 2, but use their ages 6 years from now.

$$(3x + 6) = 2(x + 6)$$

Distribute the 2.

$$3x + 6 = 2x + 12$$

Combine like terms.

$$3x - 2x = 12 - 6$$

$$x = 6$$

Susan is $x = \boxed{6}$ years old now and William is $3x = 3(6) = \boxed{18}$ years old now. Check the answers: Presently, William (18) is three times as old as Susan (6). In 6 years, Susan will be $6 + 6 = 12$ years old and William will be $18 + 6 = 24$ years old. In 6 years, William (24) will be twice as old as Susan (12).

EXAMPLE #5

Diana is 18 years older than Cara. Ten years ago, the sum of their ages was 24. What are their ages now?

SOLUTION

Define the variable clearly. Cara is younger than Diana.

$$x = \text{Cara's age now}$$

Diana is 18 years older than Cara. This involves addition.

$$x + 18 = \text{Diana's age now}$$

The second sentence refers to their ages ten years ago. Subtract 10 from their current ages to find their ages 10 years ago.

$$x - 10 = \text{Cara's age 10 years ago}$$

$$(x + 18) - 10 = x + 18 - 10 = x + 8 = \text{Diana's age 10 years ago}$$

Ten years ago, the sum of their ages was 24. Using their ages 10 years ago, their ages add up to 24.

$$(x - 10) + (x + 8) = 24$$

Combine like terms.

$$x - 10 + x + 8 = 24$$

$$2x - 2 = 24$$

$$2x = 26$$

$$x = 13$$

Cara is $x = \boxed{13}$ years old now and Diana is $x + 18 = 13 + 18 = \boxed{31}$ years old now. Check the answers: Presently, Diana (31) is 18 years older than Cara (13). Ten years ago, Cara was $13 - 10 = 3$ years old and Diana was $31 - 10 = 21$ years old. Ten years ago, their ages added up to $3 + 21 = 24$.

EXAMPLE #6

The digits of a two-digit number add up to 8. When the digits are reversed, the reversed number is smaller than the original number by 18. What is the number?

SOLUTION

Define the variable clearly. The units digit must be smaller than the tens digit because the number is smaller when the digits are reversed.

$$x = \text{the units digit of the original number}$$

The digits add up to 8. Since the units digit is x, it follows that:

$$8 - x = \text{the tens digit of the original number}$$

To make the original number, multiply the tens digit by 10 and the units digit by 1.

$$10(8 - x) + 1(x) = 80 - 10x + x = 80 - 9x = \text{the original number}$$

To make the reversed number, swap the tens digit with the units digit.

$$8 - x = \text{the units digit of the reversed number}$$

$$x = \text{the tens digit of the reversed number}$$

$$10(x) + 1(8 - x) = 10x + 8 - x = 9x + 8 = \text{the reversed number}$$

The reversed number is smaller than the original number by 18. Subtract 18 from the original number to make the reversed number.

$$(80 - 9x) - 18 = 9x + 8$$

$$80 - 9x - 18 = 9x + 8$$

$$62 = 18x + 8$$

$$54 = 18x$$

$$\frac{54}{18} = 3 = x$$

The units digit is $x = 3$ and the tens digit is $8 - x = 8 - 3 = 5$. The original number is $10(5) + 3 = 50 + 3 = \boxed{53}$. Check the answer: The reversed number is $10(3) + 5 = 35$. The original number minus the reversed number is $53 - 35 = 18$.

EXAMPLE #7

There are 350 students at a school where the ratio of girls to boys is 4:3. How many girls and how many boys attend the school?

SOLUTION

Define the variable clearly. There are fewer boys than girls.

$$x = \text{the number of boys}$$

The ratio of girls to boys is 4:3. This ratio can be expressed as the fraction $\frac{4}{3}$. Multiply the number of boys by this fraction to get the number of girls.

$$\frac{4x}{3} = \text{the number of girls}$$

The total number of students is 350. Add the number of boys to the number of girls to get the total number of students.

$$x + \frac{4x}{3} = 350$$

Would you prefer not to have fractions in the equation? If so, multiply both sides of the equation by the lowest common denominator. In this case, the lowest common denominator is 3 (it's easy when that's the only denominator in the equation). We will multiply both sides of the previous equation by 3.

$$3\left(x + \frac{4x}{3}\right) = 3(350)$$

$$3x + 4x = 1050$$

$$7x = 1050$$

$$x = \frac{1050}{7} = 150$$

There are $x = \boxed{150}$ boys and $\frac{4x}{3} = \frac{4(150)}{3} = \frac{600}{3} = \boxed{200}$ girls. Check the answers: The total number of students is $150 + 200 = 350$, and the ratio of girls to boys is $\frac{200}{150} = \frac{4}{3}$.

EXAMPLE #8

A piggy bank contains quarters, nickels, and pennies. There are eight more quarters than nickels, and there are twice as many pennies as quarters. The total value of the money in the piggy bank is $6. How many coins of each kind are in the piggy bank?

SOLUTION

Define the variable clearly. There are fewer nickels than quarters or pennies.

$$x = \text{the number of nickels}$$

There are eight more quarters than nickels.

$$x + 8 = \text{the number of quarters}$$

There are twice as many pennies as quarters.

$$2(x + 8) = 2x + 16 = \text{the number of pennies}$$

Since 1 quarter is worth 25 cents, multiply the number of quarters by 25 to find the value of the quarters. Since 1 nickel is worth 5 cents, multiply the number of nickels by 5. Since 1 penny is worth 1 cent, multiply the number of pennies by 1.

$$25(x + 8) = 25x + 200 = \text{the value of the quarters}$$

$$5(x) = 5x = \text{the value of the nickels}$$

$$1(2x + 16) = 2x + 16 = \text{the value of the pennies}$$

The total value of the piggy bank is $6. Multiply $6 by 100 to convert this to 600 cents. Add up the values of the coins and set this equal to 600 cents.

$$(25x + 200) + 5x + (2x + 16) = 600$$

$$32x + 216 = 600$$

$$32x = 600 - 216 = 384$$

$$x = \frac{384}{32} = 12$$

There are $x + 8 = 12 + 8 = \boxed{20}$ quarters, $x = \boxed{12}$ nickels, and $2(20) = \boxed{40}$ pennies. Check the answers: The total value of the coins is $25(20) + 5(12) + 1(40) = 600$.

EXAMPLE #9

A salesman sold hot dogs for \$3.25 each and sodas for \$1.75 each. The salesman sold three times as many hot dogs as sodas. The total amount paid was \$161. How many hot dogs and how many sodas did the salesman sell?

SOLUTION

Define the variable clearly. Fewer sodas were sold than hot dogs.

$$x = \text{the number of sodas sold}$$

Three times as many hot dogs were sold compared to the number of sodas sold.

$$3x = \text{the number of hot dogs sold}$$

To find the amount paid for the sodas in dollars, multiply the number of sodas (x) by 1.75. To find the amount paid for the hot dogs in dollars, multiply the number of hot dogs ($3x$) by 3.25.

$$1.75(x) = 1.75x = \text{the amount paid for the sodas}$$

$$3.25(3x) = 9.75x = \text{the amount paid for the hot dogs}$$

The total amount paid was \$161. Add the amounts paid for hot dogs and sodas.

$$1.75x + 9.75x = 161$$

Would you prefer not to have decimals in the equation? If so, multiply both sides of the equation by a number that would remove the decimal. It is usually convenient to multiply by 100 for problems that involve dollars and cents (though in this case it would also work to multiply by 4 instead). We will multiply by 100.

$$175x + 975x = 16{,}100$$

$$1150x = 16{,}100$$

$$x = \frac{16{,}100}{1150} = 14$$

The salesman sold $x = \boxed{14}$ sodas and $3x = 3(14) = \boxed{42}$ hot dogs. Check the answers: The total amount paid is $\$1.75(14) + \$3.25(42) = \$24.5 + \$136.5 = \$161$.

EXAMPLE #10

A woman pays \$480 for a laptop. The laptop was on sale for 25% off. There is no sales tax where the purchase was made. What was the regular price of the laptop?

SOLUTION

Define the variable clearly.

$$x = \text{the regular price of the laptop in dollars}$$

The laptop was on sale for 25% off. Note that saving 25% off is equivalent to paying 75%. That is, $100\% - 25\% = 75\%$. Divide 75% by 100% to convert it to a decimal: $\frac{75\%}{100\%} = 0.75$. The purchase price in dollars equals 0.75 times the regular price.

$$480 = 0.75x$$

If you multiply both sides by 100, it removes the decimal point. (Alternatively, you could write 0.75 as $\frac{3}{4}$. In that case, when you divide by $\frac{3}{4}$, it is equivalent to multiplying by $\frac{4}{3}$. If you do the math correctly, you will get the same answer either way.)

$$48{,}000 = 75x$$

$$\frac{48{,}000}{75} = 640 = x$$

The regular price of the laptop is $x = \boxed{640}$ dollars. Check the answer: 25% of \$640 is $0.25(\$640) = \$160$, such that 25% off of \$640 equals $\$640 - \$160 = \$480$.

Note: It would be incorrect to find 25% of \$480. The discount of 25% is applied to \$640, not to \$480. Compare $480 = 0.75(640) = \frac{3}{4}(640)$ with $640 = \frac{4}{3}(480)$. From one perspective, \$480 is 25% less than \$640, but from the opposite perspective, \$640 is $33\frac{1}{3}\%$ more than \$480. One calculation involves the fraction $\frac{3}{4}$ (which equals 75%); the opposite perspective involves the fraction $\frac{4}{3}$ (which equals $133\frac{1}{3}\%$).

EXAMPLE #11

A boy pays \$1.35 for a bag of candy, including 8% sales tax. What is the price of the bag of candy before tax?

SOLUTION

Define the variable clearly.

$$x = \text{the sales price in dollars before tax}$$

There is an 8% sales tax. Note that the total cost equates to 108% of the sales price: $100\% + 8\% = 108\%$ since the sales tax is in addition to the sales price. Divide 108% by 100% to convert it to a decimal: $\frac{108\%}{100\%} = 1.08$. The purchase price in dollars equals 1.08 times the sales price.

$$1.35 = 1.08x$$

If you multiply both sides by 100, it removes the decimal point.

$$135 = 108x$$

Divide both sides by 108. Note that the fraction $\frac{135}{108}$ reduces to $\frac{5}{4}$ if you divide both the numerator and denominator by 27.

$$\frac{135}{108} = \frac{5}{4} = 1.25 = x$$

The sales price of the bag of candy before tax is $x = \boxed{1.25}$ dollars. Check the answer: The sales tax is 8% of \$1.25, which works out to $0.08(\$1.25) = \$0.10$. Add the sales tax to the sales price to get the purchase price: $\$1.25 + \$0.10 = \$1.35$.

Note: It would be incorrect to find 8% of \$1.35. The sales tax of 8% is applied to \$1.25, not to \$1.35.

EXAMPLE #12

A businesswoman invested a total of $8000 in two different stocks. One stock returned 4% of its investment and the other stock returned 6% of its investment. The total return was $370. How much money did she invest in each stock?

SOLUTION

Define the variable clearly.

$$x = \text{the amount invested in the stock that paid a 4\% return}$$

The total amount invested was $8000. Part of this was invested in the stock that paid a 4% return and the remainder was invested in the stock that paid a 6% return. The two investments add up to $8000. It follows that:

$$8000 - x = \text{the amount invested in the stock that paid a 6\% return}$$

The interest earned (I) equals the principal invested (P) times the interest rate (r): $I = Pr$. Apply this formula to each investment. Divide by 100% to convert 4% and 6% into decimals: $\frac{4\%}{100\%} = 0.04$ and $\frac{6\%}{100\%} = 0.06$.

$$0.04x = \text{the return from the stock that paid 4\%}$$

$$0.06(8000 - x) = 480 - 0.06x = \text{the return from the stock that paid 6\%}$$

The total return was $370. Add the two returns together.

$$0.04x + (480 - 0.06x) = 370$$

Multiply by 100 to remove all of the decimals.

$$4x + 48{,}000 - 6x = 37{,}000$$

$$11{,}000 = 2x$$

$$\frac{11{,}000}{2} = 5500 = x$$

The woman invested $x = \boxed{5500}$ dollars in the stock that returned 4% and $8000 - 5500 = \boxed{2500}$ dollars in the stock that returned 6%. Check the answers: The total return is $0.04(\$5500) + 0.06(\$2500) = \$220 + \$150 = \$370$.

EXAMPLE #13

Two numbers have an average value of 21. One number is twice the other number. What are the numbers?

SOLUTION

Define the variable clearly.

$$x = \text{the smaller number}$$

One number is twice the other number.

$$2x = \text{the larger number}$$

The average value is 21. To find the average value of two numbers, add the numbers together and divide by two.

$$\frac{x + 2x}{2} = 21$$

Combine like terms.

$$\frac{3x}{2} = 21$$

Multiply both sides of the equation by 2.

$$3x = 2(21) = 42$$

Divide both sides of the equation by 3.

$$x = \frac{42}{3} = 14$$

The smaller number is $x = \boxed{14}$ and the larger number is $2x = 2(14) = \boxed{28}$. Check the answers: The average value is $\frac{14+28}{2} = \frac{42}{2} = 21$.

EXAMPLE #14

A dog travels 5 m/s for 100 m and then travels 8 m/s for 120 m. For how much time does the dog travel?

SOLUTION

Define the variables clearly.

$$t_1 = \text{the time for the first part of the trip}$$

$$t_2 = \text{the time for the second part of the trip}$$

Distance (d) equals rate (r) times time (t): $d = rt$. The rates are 5 m/s and 8 m/s. Apply the rate equation to each part of the trip. For the first part of the trip:

$$100 = 5t_1$$

$$\frac{100}{5} = 20 = t_1$$

For the second part of the trip:

$$120 = 8t_2$$

$$\frac{120}{8} = 15 = t_2$$

Add the two times together in order to determine the total time traveled: The dog travels for $t_1 + t_2 = 20 + 15 = \boxed{35}$ seconds. Check the answer: The distances are $5t_1 = 5(20) = 100$ and $8t_2 = 8(15) = 120$.

EXAMPLE #15

Wendy and Virginia are initially 200 m apart. At the same moment, Wendy begins traveling 4 m/s towards Virginia, and Virginia begins traveling 6 m/s towards Wendy. When will Wendy and Virginia meet?

SOLUTION

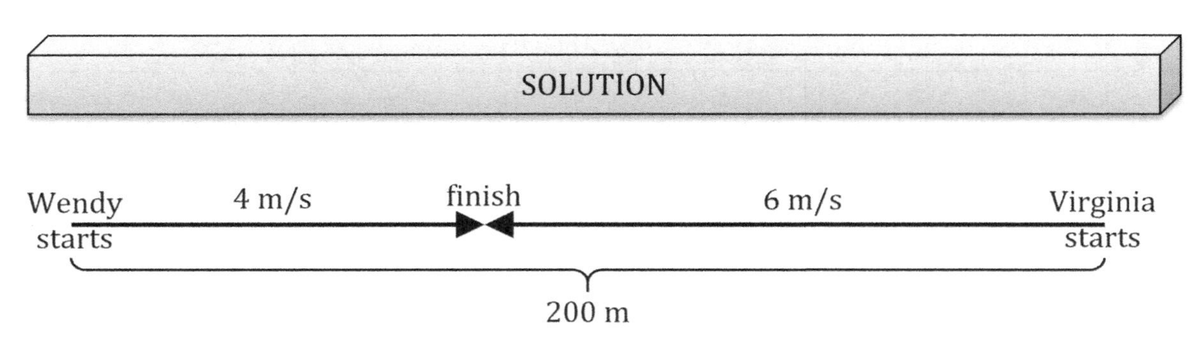

Define the variable clearly.

t = the amount of time in seconds that has passed since they began traveling

Distance (d) equals rate (r) times time (t): $d = rt$. The rates are 4 m/s and 6 m/s. Apply the rate equation to each girl. Organize the information in a table.

	rate (m/s)	time (s)	distance (m)
Wendy	4	t	$4t$
Virginia	6	t	$6t$

Wendy and Virginia are initially 200 m apart. The distances add up to 200 m.

$$4t + 6t = 200$$

$$10t = 200$$

$$t = \frac{200}{10} = 20$$

The girls travel for $t = \boxed{20}$ seconds before they meet. Check the answer: Wendy travels $4t = 4(20) = 80$ meters and Virginia travels $6t = 6(20) = 120$ meters. The total distance traveled is $80 + 120 = 200$ meters.

A father and child are initially standing together. The child begins to run away from the father with a speed of 5 ft./s. Three seconds after the child began running, the father begins to chase the child with a speed of 8 ft./s. When is the child caught?

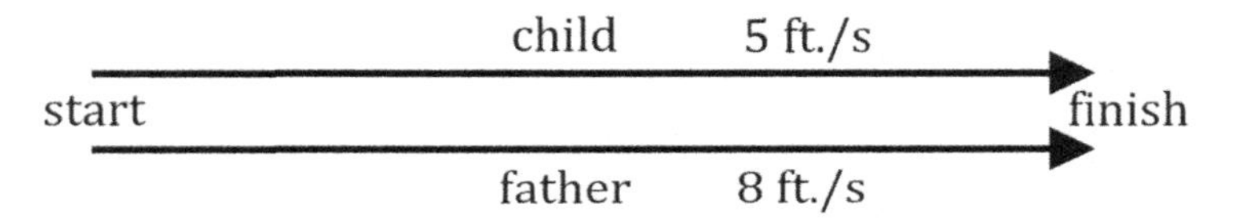

Define the variable clearly.

t = the amount of time in seconds that has passed since the child began running

The father begins running three seconds after the child began running. This means that the father spends less time running.

$t - 3$ = the amount of time that has passed since the father began running

Distance (d) equals rate (r) times time (t): $d = rt$. The rates are 5 ft./s and 8 ft./s. Apply the rate equation to each person. Organize the information in a table.

	rate (ft./s)	time (s)	distance (ft.)
child	5	t	$5t$
father	8	$t - 3$	$8(t - 3) = 8t - 24$

The father and child travel the same distance. Set their distances equal.

$$5t = 8t - 24$$
$$24 = 8t - 5t = 3t$$
$$\frac{24}{3} = 8 = t$$

The child is caught $t = \boxed{8}$ seconds after the child starts. Check the answer: The distances are $d = 5t = 5(8) = 40$ ft. and $d = 8(t - 3) = 8(8 - 3) = 8(5) = 40$ ft.

EXAMPLE #17

A motorboat travels 24 km/hr in still water. A man travels in the motorboat along a river downstream from one town to another in 30 minutes. When he returns along the same route upstream, it takes 45 minutes. What is the river current?

SOLUTION

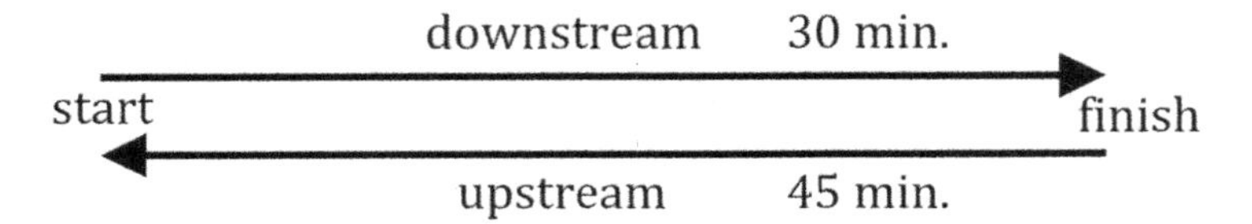

Define the variable clearly.

r = the river current in kilometers per hour (km/hr)

The motorboat travels faster downstream than upstream. Add or subtract the river current (r) to 24 km/hr to find the speeds downstream and upstream. Convert the times to hours since the rates are in km/hr: 30 min. $= \frac{1}{2}$hr and 45 min. $= \frac{3}{4}$hr.

	rate (km/hr)	time (hr)	distance (km)
downstream	$24 + r$	$\frac{1}{2}$	$(24 + r)\frac{1}{2} = 12 + \frac{r}{2}$
upstream	$24 - r$	$\frac{3}{4}$	$(24 - r)\frac{3}{4} = 18 - \frac{3r}{4}$

The boat travels the same distance each way. Set the distances equal to each other.

$$12 + \frac{r}{2} = 18 - \frac{3r}{4} \quad \text{(multiply both sides by 4)}$$

$$48 + 2r = 72 - 3r$$

$$5r = 24$$

$$r = \frac{24}{5} = 4\frac{4}{5} = 4.8$$

The river current is $r = \boxed{4.8}$ km/hr. Check the answer: The distance downstream is $(24 + 4.8)\frac{1}{2} = \frac{28.8}{2} = 14.4$ km and upstream it is $(24 - 4.8)\frac{3}{4} = \frac{57.6}{4} = 14.4$ km.

EXAMPLE #18

If 5 liters of a solution contains 1.5 liters of pure ethanol, what is the concentration of ethanol in the solution?

SOLUTION

Define the variable clearly.

$$c = \text{the concentration of ethanol in the solution}$$

The volume of pure ethanol (P) equals the decimal value of the concentration (c) times the volume of the mixture (M): $P = cM$. The volume of pure ethanol is $P = 1.5$ liters and the volume of the mixture (which is the solution) is $M = 5$ liters.

$$P = cM$$

$$1.5 = c5$$

$$\frac{1.5}{5} = \frac{3}{10} = 0.3 = c$$

Multiply by 100% to convert a decimal to a percent. The concentration of ethanol in the solution is $c = (0.3)100\% = \boxed{30\%}$. Check the answer: $P = cM = (0.3)(5) = 1.5$.

Notation: We are using P for the amount of "pure substance," M for the amount of "mixture," and c for "concentration" in decimal form. When you think of the words "pure" and "mixed," the symbols P and M make sense. However, the notation for the mixture equations isn't standard in math and science. You might see A and B, or you might see V_1 and V_2, or even a variety of other symbols instead of P and M. The beautiful thing about algebra is that it doesn't matter which symbols you use for the unknowns, provided that you are consistent and that you define your symbols clearly.

Mixture Equations

	Solution 1	Solution 2	Mixture of Solutions 1 and 2
c (decimal)	c_1	c_2	c_3
M (amount of mixture)	M_1	M_2	$M_3 = M_1 + M_2$
P (amount of pure stuff)	$P_1 = c_1 M_1$	$P_2 = c_2 M_2$	$P_3 = c_3 M_3$ $P_3 = P_1 + P_2$

Multiply row c times row M to make row P.

Add M_1 and M_2 to make M_3.

Add P_1 and P_2 to make P_3.

Example with Numbers

	Solution 1 (30%)	Solution 2 (60%)	Mixture of Solutions 1 and 2
c (decimal)	0.3	0.6	c_3 (unknown)
M (amount of mixture in liters)	10	5	$M_3 = 10 + 5 = 15$
P (amount of pure stuff)	$P_1 = (0.3)10 = 3$	$P_2 = (0.6)5 = 3$	$P_3 = 15c_3$ $P_3 = 3 + 3 = 6$

In this example:

$$15c_3 = 6$$

$$c_3 = \frac{6}{15} = 0.4 = 40\%$$

Multiply by 100% to convert 0.4 from a decimal into a percent.

EXAMPLE #19

When 6 liters of 40% hydrochloric acid is mixed with 4 liters of 20% hydrochloric acid, what percent of the resulting mixture is hydrochloric acid?

SOLUTION

Define the variable clearly.

$x =$ the percent of hydrochloric acid in the mixture, expressed as a decimal

The amount of hydrochloric acid (P) equals the decimal value of the percentage (c) times the volume of the mixture (M): $P = cM$. Apply this formula to each mixture. Divide the given percentages by 100% in order to convert them to decimals: $\frac{40\%}{100\%} = 0.4$ and $\frac{20\%}{100\%} = 0.2$. Organize the information in a table. The volume of the mixture is the sum of the volumes of the two solutions: $M_1 + M_2 = M_3 \rightarrow 6 + 4 = 10$.

	40% HCl	20% HCl	the mixture
c (decimal)	0.4	0.2	x
M (liters)	6	4	$6 + 4 = 10$
$P = cM$ (pure)	$(0.4)(6) = 2.4$	$(0.2)(4) = 0.8$	$10x$

The amount of pure stuff in the mixture is the sum of the amounts of the pure stuff in the two solutions: Add the values in the bottom row of the table.

$$2.4 + 0.8 = 10x$$

$$3.2 = 10x$$

$$\frac{3.2}{10} = 0.32 = x$$

Multiply by 100% to convert this to a percent: $x = (0.32)100\% = \boxed{32\%}$. Check the answer: $P_3 = c_3M_3 = (0.32)(10) = 3.2$ agrees with $P_1 + P_2 = 2.4 + 0.8 = 3.2 = P_3$.

EXAMPLE #20

A 10% ethanol solution is mixed with a 25% ethanol solution. How much of each solution should be combined in order to make a 12-gallon mixture that is 15% ethanol?

SOLUTION

Define the variable clearly.

$$x = \text{the volume in gallons of 10\% ethanol solution}$$

The amount of ethanol (P) equals the decimal value of the percentage (c) times the volume of the solution (M): $P = cM$. Apply this formula to each mixture. Organize the information in a table. The volume of the mixture is the sum of the volumes of the two solutions: $M_1 + M_2 = M_3 \rightarrow M_2 = 12 - x$.

	10% ethanol	25% ethanol	15% ethanol
c (decimal)	0.1	0.25	0.15
M (gallons)	x	$12 - x$	12
$P = cM$ (pure)	$0.1x$	$0.25(12 - x)$	$(0.15)(12) = 1.8$

The amount of pure stuff in the mixture is the sum of the amounts of the pure stuff in the two solutions: Add the values in the bottom row of the table.

$$0.1x + 0.25(12 - x) = 1.8$$

$$0.1x + 3 - 0.25x = 1.8 \quad \text{(multiply both sides by 100)}$$

$$10x + 300 - 25x = 180$$

$$120 = 15x$$

$$\frac{120}{15} = 8 = x$$

The volumes are $x = \boxed{8}$ gallons of 10% ethanol and $12 - x = 12 - 8 = \boxed{4}$ gallons of 25% ethanol. Check the answers: $P_1 + P_2 = 0.1(8) + 0.25(4) = 0.8 + 1 = 1.8 = P_3$.

EXAMPLE #21

Approximately, how much water must be added to 80 cc of a 35% antifreeze solution in order to dilute the antifreeze solution down to 25%?

SOLUTION

Define the variable clearly.

$x =$ the volume of water in cubic centimeters that needs to be added

The amount of pure antifreeze (P) equals the decimal value of the percentage (c) times the volume of the mixture (M): $P = cM$. Apply this formula to each mixture. Divide the given percentages by 100% in order to convert them to decimals: $\frac{35\%}{100\%} = 0.35$ and $\frac{25\%}{100\%} = 0.25$. Organize the information in a table. The volume of the mixture is the sum of the volumes of the two solutions: $M_1 + M_2 = M_3$. Water is 0% antifreeze.

	35% antifreeze	water (0% antifreeze)	25% antifreeze
c (decimal)	0.35	0	0.25
M (cc)	80	x	$80 + x$
$P = cM$ (pure)	$(0.35)(80) = 28$	$0x = 0$	$0.25(80 + x)$

The amount of pure stuff in the mixture is the sum of the amounts of the pure stuff in the two solutions: Add the values in the bottom row of the table.

$$28 + 0 = 0.25(80 + x)$$
$$28 = 20 + 0.25x$$
$$8 = 0.25x$$
$$\frac{8}{0.25} = 32 = x$$

The volume of water needed is $x = \boxed{32}$ cc. Check the answer: $P_1 + P_2 = 28 + 0 = 28$ agrees with $P_3 = c_3 M_3 = 0.25(80 + x) = 0.25(80 + 32) = 0.25(112) = 28$.

EXAMPLE #22

Nancy could paint a house in 6 hours. Larry could paint the same house in 8 hours. If they work together to paint the house, about how long would it take?

SOLUTION

Define the variable clearly.

$t =$ the time it takes Nancy and Larry to paint the house together

Find the reciprocal of each given time in order to determine how much work each person could complete in one hour.

	Nancy	Larry	together
time to work (in hours)	6	8	t
fraction completed per hour	$\frac{1}{6}$	$\frac{1}{8}$	$\frac{1}{t}$

When people work together, add the reciprocals from the bottom row of the table.

$$\frac{1}{6} + \frac{1}{8} = \frac{1}{t}$$

The lowest common denominator of 6, 8, and t is $24t$. Multiply both sides by $24t$.

$$\frac{24t}{6} + \frac{24t}{8} = \frac{24t}{t}$$

$$4t + 3t = 24$$

$$7t = 24$$

$$t = \frac{24}{7} = 3\frac{3}{7}$$

Nancy and Larry could paint the house in about $t = \boxed{3\frac{3}{7}} \approx \boxed{3.4}$ hours if they work together. Check the answer: The fraction of the work that each person does is $3.4 \div 6$ and $3.4 \div 8$. These should approximately add up to one: $3.4 \div 6 + 3.4 \div 8 \approx 1$.

EXAMPLE #23

Paul could vacuum a house in 18 minutes by himself. If Paul and Kelly work together, it takes them 12 minutes to vacuum the house. How long would it take for Kelly to vacuum the house by himself? (Paul and Kelly each have their own vacuum cleaner.)

SOLUTION

Define the variable clearly.

$t =$ the time it takes Kelly to vacuum the house by himself

Find the reciprocal of each given time in order to determine how much work each person could complete in one minute.

	Paul	Kelly	together
time to work (in minutes)	18	t	12
fraction completed per minute	$\frac{1}{18}$	$\frac{1}{t}$	$\frac{1}{12}$

When people work together, add the reciprocals from the bottom row of the table.

$$\frac{1}{18} + \frac{1}{t} = \frac{1}{12}$$

The lowest common denominator of 18, t, and 12 is $36t$. Multiply both sides by $36t$.

$$\frac{36t}{18} + \frac{36t}{t} = \frac{36t}{12}$$

$$2t + 36 = 3t$$

$$36 = t$$

Kelly could vacuum the house by himself in about $t = \boxed{36}$ minutes. Check the answer: The fraction of the work that each person does is $\frac{12}{18}$ and $\frac{12}{36}$. (Divide the time it takes to work together, 12 minutes, by each time it takes to work alone.) These fractions should add up to one: $\frac{12}{18} + \frac{12}{36} = \frac{2}{3} + \frac{1}{3} = \frac{3}{3} = 1$.

EXAMPLE #24

An empty sink could fill up in 3 minutes when the drain is plugged. When the same sink is full, it could drain completely in 4 minutes. If the sink is empty to begin with and the drain is unplugged, how long would it take for the sink to fill up?

SOLUTION

Define the variable clearly.

t = the time it takes the sink to fill up when the drain is unplugged

Find the reciprocal of each given time in order to determine how much work is done each minute.

	fill up (plugged)	drain	fill up (unplugged)
time to work (in minutes)	3	4	t
fraction completed per minute	$\frac{1}{3}$	$\frac{1}{4}$	$\frac{1}{t}$

Since the faucet and drain work against each other (instead of working together), we subtract the reciprocals from the bottom row of the table (instead of adding).

$$\frac{1}{3} - \frac{1}{4} = \frac{1}{t}$$

The lowest common denominator of 3, 4, and t is $12t$. Multiply both sides by $12t$.

$$\frac{12t}{3} - \frac{12t}{4} = \frac{12t}{t}$$

$$4t - 3t = 12$$

$$t = 12$$

The sink will take $t = \boxed{12}$ minutes to fill up when the drain is unplugged. Check the answer: The fraction of the work done by the faucet and drain is $\frac{12}{3}$ and $\frac{12}{4}$. Since the drain works against the faucet, subtract these to make one: $\frac{12}{3} - \frac{12}{4} = 4 - 3 = 1$.

EXAMPLE #25

If the system below is in static equilibrium, what is the value of r_2?

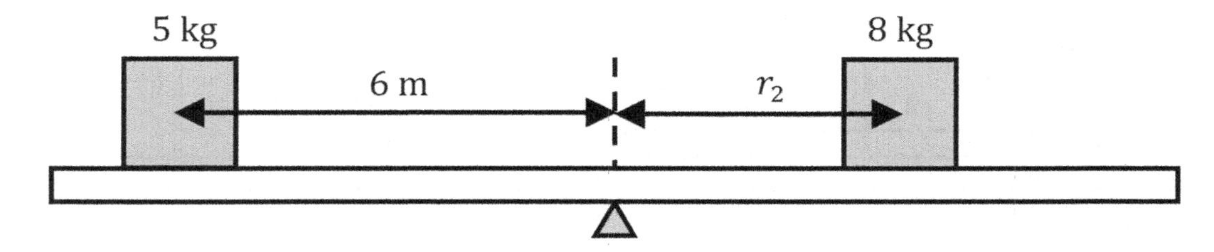

SOLUTION

Define the variable clearly.

r_2 = the distance from the fulcrum to the center of the 8-kg box

When a lever is in static equilibrium, the counterclockwise torques and clockwise torques are equal. (Note that the weights are <u>**not**</u> equal.) Set the sum of the counter-clockwise torques equal to the sum of the clockwise torques, where each torque equals weight times lever arm. (See the side note below regarding weight.)

$$w_1 r_1 = w_2 r_2$$
$$(5)(6) = (8)(r_2)$$
$$30 = 8r_2$$
$$\frac{30}{8} = \frac{15}{4} = 3\frac{3}{4} = 3.75 = r_2$$

The 8-kg box is $r_2 = \boxed{3\frac{3}{4}} = \boxed{3.75}$ meters from the fulcrum. Check the answer: The counterclockwise torque is $(5)(6) = 30$ and the clockwise torque is $(8)(3.75) = 30$.

Side notes: Strictly speaking, weight equals mass times gravity (9.81 m/s^2). However, since every term in the equation includes weight, gravity cancels out. This means that we may choose to work with mass instead of weight (so there is no reason to multiply all of the masses by 9.81 m/s^2). Also, the diagram is <u>not</u> drawn to scale.

EXAMPLE #26

If the system below is in static equilibrium, what is the value of w_3?

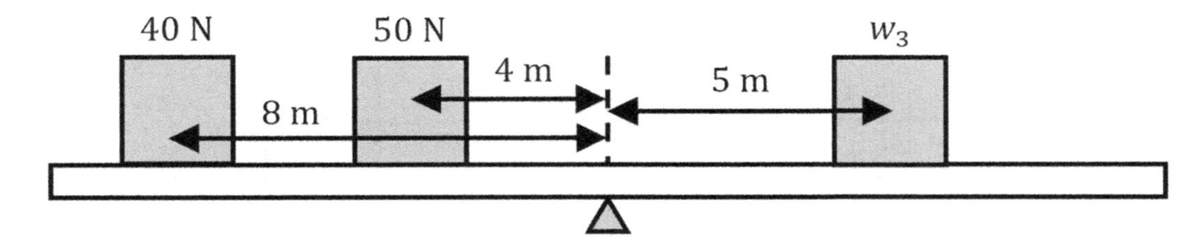

SOLUTION

Define the variable clearly.

$$w_3 = \text{the weight of the box on the right side of the fulcrum}$$

When a lever is in static equilibrium, the counterclockwise torques and clockwise torques are equal. (Note that the weights are **not** equal.) Set the sum of the counterclockwise torques equal to the sum of the clockwise torques, where each torque equals weight times lever arm.

$$w_1 r_1 + w_2 r_2 = w_3 r_3$$
$$(40)(8) + (50)(4) = (w_3)(5)$$
$$320 + 200 = 5w_3$$
$$520 = 5w_3$$
$$\frac{520}{5} = 104 = w_3$$

The box on the right side of the fulcrum has a weight of $w_3 = \boxed{104}$ Newtons. Check the answer: The counterclockwise torque is $(40)(8) + (50)(4) = 320 + 200 = 520$ and the clockwise torque is $(104)(5) = 520$.

Side note: Unlike the previous example, this example works with weight in Newtons instead of mass in kilograms. The only difference the units make for this example is what units the answer for w_3 has.

EXAMPLE #27

The system below is in static equilibrium. The two boxes have a combined weight of 200 lbs. How much does each box weigh?

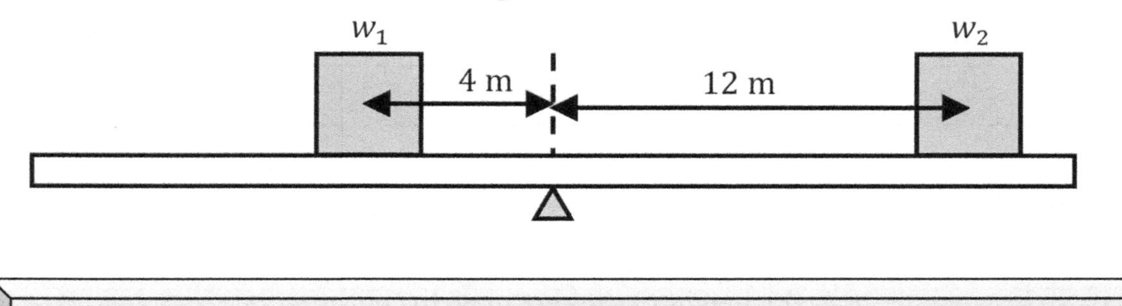

SOLUTION

Define the variables clearly.

$$w_1 = \text{the weight of the box on the left side of the fulcrum}$$
$$w_2 = \text{the weight of the box on the right side of the fulcrum}$$

The system is in static equilibrium. Apply the torque equation.

$$w_1 r_1 = w_2 r_2$$
$$4w_1 = 12w_2 \quad \text{(first equation)}$$

The combined weight of the boxes is 200 lbs.

$$w_1 + w_2 = 200 \quad \text{(second equation)}$$

Isolate one of the unknowns. Divide both sides of the first equation by 4.

$$w_1 = 3w_2$$

Substitute this expression in for w_1 into the second equation.

$$3w_2 + w_2 = 200$$
$$4w_2 = 200$$
$$w_2 = \frac{200}{4} = 50$$

The box on the right side of the fulcrum has a weight of $w_2 = \boxed{50}$ lbs. and the box on the left side of the fulcrum has a weight of $w_1 = 3w_2 = 3(50) = \boxed{150}$ lbs. Check the answers: The two torques are $4(150) = 600$ and $(12)(50) = 600$.

EXAMPLE #28

A rectangle has a perimeter of 40 cm and an area of 96 cm². What are the width and height of the rectangle?

SOLUTION

Define the variables clearly.

$$w = \text{the width of the rectangle in centimeters}$$
$$h = \text{the height of the rectangle in centimeters}$$

Plug the given area and perimeter into the formulas for the area and perimeter of a rectangle, which are $A = wh$ and $P = 2w + 2h$.

$$96 = wh \quad \text{(area equation)}$$
$$40 = 2w + 2h \quad \text{(perimeter equation)}$$

Isolate h in the area equation. Divide both sides by w.

$$\frac{96}{w} = h$$

Substitute this expression for h into the perimeter equation.

$$40 = 2w + 2\left(\frac{96}{w}\right) \quad \text{(multiply both sides by } w\text{)}$$
$$40w = 2w^2 + 192 \quad \text{(rearrange the terms)}$$
$$-2w^2 + 40w - 192 = 0 \quad \text{(apply the quadratic formula)}$$
$$w = \frac{-40 \pm \sqrt{40^2 - 4(-2)(-192)}}{2(-2)} = \frac{-40 \pm \sqrt{1600 - 1536}}{-4}$$
$$w = \frac{-40 \pm \sqrt{64}}{-4} = \frac{-40 \pm 8}{-4} = \frac{-48}{-4} \text{ or } \frac{-32}{-4} = 12 \text{ or } 8$$

The width of the rectangle is $w = \boxed{12}$ cm and the height is $h = \frac{96}{w} = \frac{96}{12} = \boxed{8}$ cm. Check the answers: The perimeter is $P = 2w + 2h = 2(12) + 2(8) = 24 + 16 = 40$ cm and the area is $A = wh = (12)(8) = 96$ cm^2.

EXAMPLE #29

The base of a triangle is twice its height. The area of the triangle is 9 in.^{2} What are the base and height of the triangle?

SOLUTION

Define the variable clearly.

$$b = \text{the base of the triangle in inches}$$
$$h = \text{the height of the triangle in inches}$$

Plug the given area into the formula for the area of a triangle, which is $A = \frac{1}{2}bh$.

$$9 = \frac{1}{2}bh \quad \text{(area equation)}$$

The base is twice the height.

$$b = 2h \quad \text{(base equation)}$$

Substitute this expression for the base into the area equation.

$$9 = \frac{1}{2}(2h)h$$
$$9 = h^2$$

Squareroot both sides of the equation.

$$\sqrt{9} = h$$
$$3 = h$$

The height of the triangle is $h = \boxed{3}$ in. and the base is $b = 2h = 2(3) = \boxed{6}$ in. Check the answers: The area is $A = \frac{1}{2}bh = \frac{1}{2}(6)(3) = 9$ in.^{2}

EXAMPLE #30

A circle has a diameter of 6 cm. What is the area of the circle?

SOLUTION

Define the variables clearly.

$$r = \text{the radius of the circle in centimeters}$$

$$A = \text{the area of the circle in square centimeters}$$

Use the formula for the diameter of a circle to determine the radius.

$$D = 2r$$

$$6 = 2r$$

$$\frac{6}{2} = 3 = r$$

Now use the formula for the area of a circle.

$$A = \pi r^2 = \pi(3)^2 = 9\pi \approx 9(3.14159) \approx 28$$

The area of the circle is $A = 9\pi \approx \boxed{28}$ cm^2. Check the answer: Rewrite $D = 2r$ as $r = \frac{D}{2}$ and substitute this into $A = \pi r^2$ to get $A = \pi\left(\frac{D}{2}\right)^2 = \frac{\pi D^2}{4} = \frac{\pi(6)^2}{4} = \frac{36\pi}{4} = 9\pi$.

WORD PROBLEMS

The problems start out easy and grow in difficulty.

The first several problems are very similar to the examples.

Later in the book, the problems will start to differ from the examples.

The same general strategies apply to the problems as they do to the examples.

When the problems become somewhat different from the examples,

you may need to think your way through the problems more.

First try to solve each problem on your own.

If you need help, review Chapters 1-3.

Try to find similar examples in Chapter 3 to help guide your reasoning.

For the harder problems, it won't be sufficient to merely copy an example.

Rather, you will need to try to understand the logic and reasoning involved.

Approach each problem with self-confidence and determination.

Students often become good problem-solvers by being confident that

the solution exists and by being very determined to figure out the answer.

Try not to give up. Think you can. Think you can. Yes, you can.

ANSWERS

On the page following each problem,

you can find the full solution.

When you are ready to check your answer,

simply turn the page and read the solution.

Be careful not to turn the page until

you are ready to check your answer.

If your answer turns out to be wrong,

try to learn from your mistakes.

Students often become good problem-solvers

by learning to avoid mistake that they have made in the past.

The end of each solution will show you

how to check the answer to the problem.

Being able to check your answer for

consistency is a valuable skill.

Try to learn how to check your own answer

before you check the solution on the back of the page.

Believe in yourself. You can do it.

PROBLEM #1

One number is ten times another number. The numbers have a sum of 825. What are the numbers?

SOLUTION #1

$$x = \text{the smaller number}$$

The larger number is ten times the smaller number.

$$10x = \text{the larger number}$$

The sum equals 825. The two numbers add up to 825.

$$x + 10x = 825$$

$$11x = 825$$

$$x = \frac{825}{11} = 75$$

The smaller number is $x = \boxed{75}$ and the larger number equals $10x = 10(75) = \boxed{750}$.

Check the answers: $75 + 750 = 825$ and $75(10) = 750$.

PROBLEM #2

Four consecutive even numbers add up to 244. What are the numbers?

SOLUTION #2

$$x = \text{the smallest number}$$

Consecutive even numbers have a difference of 2. Since the smaller number is x, it follows that:

$$x + 2 = \text{the second number}$$

$$x + 4 = \text{the third number}$$

$$x + 6 = \text{the largest number}$$

The four numbers add up to 244. Add the four numbers together.

$$x + (x + 2) + (x + 4) + (x + 6) = 244$$

Isolate the unknown. Combine like terms.

$$x + x + 2 + x + 4 + x + 6 = 244$$

$$4x + 12 = 244$$

$$4x = 244 - 12$$

$$4x = 232$$

$$x = \frac{232}{4} = 58$$

The first number is $x = \boxed{58}$, the second number is $x + 2 = 58 + 2 = \boxed{60}$, the third number is $x + 4 = 58 + 4 = \boxed{62}$, and the last number is $x + 6 = 58 + 6 = \boxed{64}$. Check the answers: 58, 60, 62, and 64 are consecutive even numbers and their sum is $58 + 60 + 62 + 64 = 244$.

PROBLEM #3

Tracy is 16 years younger than Sarah. Twelve years from now, Sarah will be twice as old as Tracy. What are their ages now?

SOLUTION #3

Tracy is younger than Sarah.

$$x = \text{Tracy's age now}$$

Tracy is 16 years younger than Sarah. Therefore, Sarah is 16 years older than Tracy.

$$x + 16 = \text{Sarah's age now}$$

The second sentence refers to their ages 12 years from now. Add 12 to their current ages to find their ages 12 years from now.

$$x + 12 = \text{Tracy's age 12 years from now}$$

$$(x + 16) + 12 = x + 28 = \text{Sarah's age 12 years from now}$$

12 years from now, Sarah will be twice as old as Tracy. Multiply Tracy's age by 2, but use their ages 12 years from now.

$$(x + 28) = 2(x + 12)$$

Distribute the 2.

$$x + 28 = 2x + 24$$

Combine like terms.

$$28 - 24 = 2x - x$$

$$4 = x$$

Tracy is $x = \boxed{4}$ years old now and Sarah is $x + 16 = 4 + 16 = \boxed{20}$ years old now. Check the answers: Presently, Tracy (4) is 16 years younger than Sarah (20). In 12 years, Tracy will be $4 + 12 = 16$ years old and Sarah will be $20 + 12 = 32$ years old. In 12 years, Sarah (32) will be twice as old as Tracy (16).

PROBLEM #4

The sum of the digits of a two-digit number is 11. When the digits are reversed, the reversed number is larger than the original number by 45. What is the number?

SOLUTION #4

The tens digit must be smaller than the units digit because the number is larger when the digits are reversed.

$$x = \text{the tens digit of the original number}$$

The sum of the digits is 11. Since the tens digit is x, it follows that:

$$11 - x = \text{the units digit of the original number}$$

To make the original number, multiply the tens digit by 10 and the units digit by 1.

$$10(x) + 1(11 - x) = 10x + 11 - x = 9x + 11 = \text{the original number}$$

To make the reversed number, swap the tens digit with the units digit.

$$11 - x = \text{the tens digit of the reversed number}$$

$$x = \text{the units digit of the reversed number}$$

$$10(11 - x) + 1(x) = 110 - 10x + x = 110 - 9x = \text{the reversed number}$$

The reversed number is larger than the original number by 45. Add 45 to the original number to make the reversed number.

$$(9x + 11) + 45 = 110 - 9x$$

$$9x + 11 + 45 = 110 - 9x$$

Add $9x$ to both sides. Note that $9x$ doesn't cancel: Instead, you get $9x + 9x = 18x$.

$$18x + 56 = 110$$

$$18x = 110 - 56$$

$$18x = 54$$

$$x = \frac{54}{18} = 3$$

The tens digit is $x = 3$ and the units digit equals $11 - x = 11 - 3 = 8$. The original number is $10(3) + 8 = 30 + 8 = \boxed{38}$. Check the answer: The reversed number is $10(8) + 3 = 83$. The original number is related to the reversed number by $38 + 45 = 83$.

PROBLEM #5

At a large gathering, the ratio of children to adults is 5:3. There are 50 more children than adults. How many children and how many adults are there?

SOLUTION #5

There are fewer adults than children.

$$x = \text{the number of adults}$$

The ratio of children to adults is 5:3. This ratio can be expressed as the fraction $\frac{5}{3}$. Multiply the number of adults by this fraction to get the number of children. (Note: If instead you defined the variable as the number of children, then you need to multiply by $\frac{3}{5}$ to get the number of adults in the next step.)

$$\frac{5x}{3} = \text{the number of children}$$

There are 50 more children than adults. Add 50 to the number of adults to get the number of children.

$$\frac{5x}{3} = x + 50$$

Multiply both sides of the equation by 3. This eliminates the denominator.

$$5x = 3(x + 50)$$

$$5x = 3x + 150$$

$$2x = 150$$

$$x = \frac{150}{2} = 75$$

There are $x = \boxed{75}$ adults and $\frac{5x}{3} = \frac{5(75)}{3} = \frac{375}{3} = \boxed{125}$ children. Check the answers: There are $125 - 75 = 50$ more children than adults, and the ratio of children to adults is $\frac{125}{75} = \frac{125 \div 25}{75 \div 25} = \frac{5}{3}$.

PROBLEM #6

A computer store sold half as many desktops as laptops. The desktops sold for $400 each and the laptops sold for $600 each. There is no sales tax where the purchases were made. The total amount paid was $48,000. How many desktops and how many laptops did the store sell?

SOLUTION #6

Fewer desktops were sold than laptops.

$$x = \text{the number of desktops sold}$$

Half as many desktops were sold compared to the number of laptops sold. This means that twice as many laptops were sold as desktops. (If you define x as the number of laptops, then the number of desktops would be $\frac{x}{2}$. If you do the math correctly, you will get the same two answers, except that x would be the number of laptops instead of the number of desktops.)

$$2x = \text{the number of laptops sold}$$

To find the amount paid for the desktops in dollars, multiply the number of desktops (x) by 400. To find the amount paid for the laptops in dollars, multiply the number of laptops ($2x$) by 600.

$$400(x) = 400x = \text{the amount paid for the desktops}$$

$$600(2x) = 1200x = \text{the amount paid for the laptops}$$

The total amount paid was \$48,000. Add the amounts paid for desktops and laptops.

$$400x + 1200x = 48{,}000$$

$$1600x = 48{,}000$$

$$x = \frac{48{,}000}{1600} = 30$$

The store sold $x = \boxed{30}$ desktops and $2x = 2(30) = \boxed{60}$ laptops. Check the answers: The total amount paid is $\$400(30) + \$600(60) = \$12{,}000 + \$36{,}000 = \$48{,}000$.

PROBLEM #7

A girl pays $4.77 for her lunch, including 6% sales tax. What is the price of the lunch before tax?

SOLUTION #7

$$x = \text{the price in dollars of the lunch before tax}$$

There is a 6% sales tax. Note that the total cost equates to 106% of the sales price: $100\% + 6\% = 106\%$ since the sales tax is in addition to the sales price. Divide 106% by 100% to convert it to a decimal: $\frac{106\%}{100\%} = 1.06$. The purchase price in dollars equals 1.06 times the sales price.

$$4.77 = 1.06x$$

Multiply both sides of the equation by 100 in order to remove the decimal point.

$$477 = 106x$$

Divide both sides by 106.

$$\frac{477}{106} = \frac{477 \div 53}{106 \div 53} = \frac{9}{2} = 4.5 = x$$

The price of the lunch before tax is $x = \boxed{4.50}$ dollars. Check the answer: The sales tax is 6% of \$4.50, which works out to $0.06(\$4.50) = \$0.27$. Add the sales tax to the sales price to get the purchase price: $\$4.50 + \$0.27 = \$4.77$.

Note: It would be incorrect to find 6% of \$4.77. The sales tax of 6% is applied to \$4.50, not to \$4.77.

PROBLEM #8

The total value of coins in a purse is $44. There are twice as many dimes as nickels, 16 fewer quarters than nickels, and five times as many pennies as dimes. How many coins of each kind are in the purse?

SOLUTION #8

There are fewer quarters than pennies, nickels, or dimes.

$$x = \text{the number of quarters}$$

There are 16 fewer quarters than nickels. This means that there are 16 more nickels than quarters.

$$x + 16 = \text{the number of nickels}$$

There are twice as many dimes as nickels.

$$2(x + 16) = 2x + 32 = \text{the number of dimes}$$

There are five times as many pennies as dimes.

$$5(2x + 32) = 10x + 160 = \text{the number of pennies}$$

Since 1 quarter is worth 25 cents, multiply the number of quarters by 25 to find the value of the quarters. Since 1 dime is worth 10 cents, multiply the number of dimes by 10. Since 1 nickel is worth 5 cents, multiply the number of nickels by 5. Since 1 penny is worth 1 cent, multiply the number of pennies by 1.

$$25(x) = 25x = \text{the value of the quarters}$$

$$10(2x + 32) = 20x + 320 = \text{the value of the dimes}$$

$$5(x + 16) = 5x + 80 = \text{the value of the nickels}$$

$$1(10x + 160) = 10x + 160 = \text{the value of the pennies}$$

The total value of the coins is \$44. Multiply \$44 by 100 to convert this to 4400 cents. Add up the values of the coins and set this equal to 4400 cents.

$$25x + (20x + 320) + (5x + 80) + (10x + 160) = 4400$$

$$60x + 560 = 4400$$

$$60x = 4400 - 560 = 3840$$

$$x = \frac{3840}{60} = 64$$

There are $x = \boxed{64}$ quarters, $2x + 32 = 2(64) + 32 = 128 + 32 = \boxed{160}$ dimes, $x + 16 = 64 + 16 = \boxed{80}$ nickels, and $10x + 160 = 10(64) + 160 = 640 + 160 = \boxed{800}$ pennies. Check the answers: The total value of the coins is $25(64) + 10(160) + 5(80) + 800(1) = 1600 + 1600 + 400 + 800 = 4400$.

PROBLEM #9

A company invested $50,000. Part of the investment was put into a savings account that paid 3% interest, and the remainder of the investment was put into stocks that paid a return of 9%. The total return was $2700. How much money was put into the savings account and how much money was invested in stocks?

SOLUTION #9

$x =$ the amount invested in the savings account that paid a 3% return

The total amount invested was \$50,000. Part of this was put into a savings account and the remainder was invested in stocks. The two investments add up to \$50,000. It follows that:

$50{,}000 - x =$ the amount invested in the stocks that paid a 9% return

The interest earned (I) equals the principal invested (P) times the interest rate (r): $I = Pr$. Apply this formula to each investment. Divide by 100% to convert 3% and 9% into decimals: $\frac{3\%}{100\%} = 0.03$ and $\frac{9\%}{100\%} = 0.09$.

$0.03x =$ the interest earned from the savings account that paid 3%

$0.09(50{,}000 - x) = 4500 - 0.09x =$ the return from the stocks that paid 9%

The total return was \$2700. Add the two returns together.

$$0.03x + (4500 - 0.09x) = 2700$$

Multiply by 100 to remove all of the decimals.

$$3x + 450{,}000 - 9x = 270{,}000$$

$$180{,}000 = 6x$$

$$\frac{180{,}000}{6} = 30{,}000 = x$$

The company invested $x = \boxed{30{,}000}$ dollars in a savings account that paid 3% interest and $50{,}000 - 30{,}000 = \boxed{20{,}000}$ dollars in stocks that paid a return of 9%. Check the answers: The total return equals $0.03(\$30{,}000) + 0.09(\$20{,}000) = \$900 + \$1800 = \$2700$.

PROBLEM #10

Two numbers have an average value of 283 and a difference of 56. What are the numbers?

SOLUTION #10

$$x = \text{the smaller number}$$

The numbers have a difference of 56. This means that one number is 56 more than the other number. It follows that:

$$x + 56 = \text{the larger number}$$

The average value is 283. To find the average value of two numbers, add the numbers together and divide by two. (In general, divide by how many numbers you are adding together. In this case, we are adding two numbers, so we divide by two.)

$$\frac{x + (x + 56)}{2} = 283$$

Combine like terms.

$$\frac{2x + 56}{2} = 283$$

Multiply both sides of the equation by 2.

$$2x + 56 = 566$$

$$2x = 510$$

$$x = \frac{510}{2} = 255$$

The smaller number is $x = \boxed{255}$ and the larger number is $x + 56 = 255 + 56 = \boxed{311}$.

Check the answers: The average value is $\frac{255+311}{2} = \frac{566}{2} = 283$.

PROBLEM #11

Kevin and Carly are initially 40 km apart. At the same moment, Kevin rides a bicycle 20 km/hr towards Carly, and Carly rides a bicycle 30 km/hr towards Kevin. When and where do Kevin and Carly meet?

SOLUTION #11

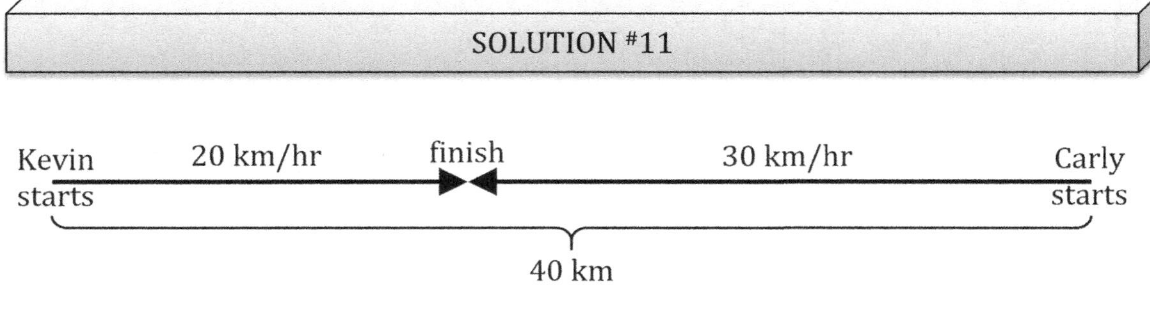

$t =$ the amount of time in hours that has passed since they began traveling

$d_1 =$ the distance in kilometers that Kevin travels before he meets Carly

Distance (d) equals rate (r) times time (t): $d = rt$. The rates are 20 km/hr and 30 km/hr. Apply the rate equation to each person. Organize the information in a table.

	rate (km/hr)	time (hr)	distance (km)
Kevin	20	t	$20t$
Carly	30	t	$30t$

Kevin and Carly are initially 40 km apart. The distances add up to 40 km.

$$20t + 30t = 40$$

$$50t = 40$$

$$t = \frac{40}{50} = \frac{4}{5} = 0.8$$

This is only one of the two answers. We also need to determine where they meet. Plug this time into the rate equation for Kevin to determine how far Kevin travels.

$$d_1 = r_1 t = (20)(0.8) = 16$$

Kevin and Carly travel for $t = \boxed{0.8}$ hours before they meet, and they meet $d_1 = \boxed{16}$ kilometers from Kevin's starting position. If you would prefer to express the time in minutes, the time is $t = 48$ minutes since 1 hour equals 60 minutes and $(0.8)(60) = 48$. If you would prefer to determine the distance Carly travels, her distance is $d_2 = r_2 t = (30)(0.8) = 24$ kilometers. Check the answers: The total distance traveled equals $16 + 24 = 40$ kilometers.

PROBLEM #12

A solution contains 6 quarts of 40% ethanol. (The concentration was measured by comparing the volume of ethanol to the volume of the solution.) What is the volume of pure ethanol contained in the solution?

SOLUTION #12

$P =$ the volume of pure ethanol contained in the solution

The volume of pure ethanol (P) equals the decimal value of the concentration (c) times the volume of the mixture (M): $P = cM$. The volume of the mixture (which is the solution) is $M = 6$ quarts. Divide by 100% to convert the concentration from a percent to a decimal: $c = \frac{40\%}{100\%} = 0.4$.

$$P = cM$$
$$P = (0.4)(6) = 2.4$$

The solution contains $P = \boxed{2.4}$ quarts of pure ethanol. Check the answer: $c = \frac{P}{M} = \frac{2.4}{6} = 0.4 = 40\%$.

(Why is there a note in parentheses in the problem? It wasn't intended to confuse you. The reason for the note is that in chemistry there are multiple ways of measuring concentration. For example, concentration is often expressed as the mass of the pure substance divided by the volume of the solution, but when a solution contains two liquids the concentration is sometimes expressed as the volume of the pure substance divided by the volume of the solution. You would have no way of knowing how the concentration was measured unless the problem stated this. That's why this problem included a note in parentheses. Don't worry: You won't need to know any chemistry to solve the problems in this book. You just need to be able to tell which substance is the pure substance and which is the mixture. You won't need to worry about whether to express the amount of pure substance as a mass or a volume. Just be consistent with the units that are given. For example, in this problem, the amounts are volumes in quarts.)

PROBLEM #13

When 200 mL of 25% sulfuric acid is mixed with 400 mL of 10% sulfuric acid, what percent of the resulting mixture is sulfuric acid?

SOLUTION #13

$x =$ the percent of sulfuric acid in the mixture, expressed as a decimal

The amount of sulfuric acid (P) equals the decimal value of the percentage (c) times the volume of the mixture (M): $P = cM$. Apply this formula to each mixture. Divide the given percentages by 100% in order to convert them to decimals: $\frac{25\%}{100\%} = 0.25$ and $\frac{10\%}{100\%} = 0.1$. Organize the information in a table. The volume of the mixture is the sum of the volumes of the two solutions: $M_1 + M_2 = M_3 \rightarrow 200 + 400 = 600$.

	25% H_2SO_4	10% H_2SO_4	the mixture
c (decimal)	0.25	0.1	x
M (mL)	200	400	$200 + 400 = 600$
$P = cM$ (pure)	$(0.25)(200) = 50$	$(0.1)(400) = 40$	$600x$

The amount of pure stuff in the mixture is the sum of the amounts of the pure stuff in the two solutions: Add the values in the bottom row of the table.

$$50 + 40 = 600x$$

$$90 = 600x$$

$$x = \frac{90}{600} = \frac{3}{20} = 0.15$$

Multiply by 100% to convert this to a percent: $x = (0.15)100\% = \boxed{15\%}$. Check the answer: $P_3 = c_3M_3 = (0.15)(600) = 90$ agrees with $P_1 + P_2 = 50 + 40 = 90 = P_3$.

PROBLEM #14

Pete could wallpaper a room in 3 hours. Brenda could wallpaper the same room in 2 hours. If they work together to wallpaper the room, about how long would it take?

SOLUTION #14

$t =$ the time it takes Pete and Brenda to wallpaper the house together

Find the reciprocal of each given time in order to determine how much work each person could complete in one hour.

	Pete	Brenda	together
time to work (in hours)	3	2	t
fraction completed per hour	$\frac{1}{3}$	$\frac{1}{2}$	$\frac{1}{t}$

When people work together, add the reciprocals from the bottom row of the table.

$$\frac{1}{3} + \frac{1}{2} = \frac{1}{t}$$

The lowest common denominator of 3, 2, and t is $6t$. Multiply both sides by $6t$.

$$\frac{6t}{3} + \frac{6t}{2} = \frac{6t}{t}$$

$$2t + 3t = 6$$

$$5t = 6$$

$$t = \frac{6}{5} = 1\frac{1}{5} = 1.2$$

Pete and Brenda could wallpaper the room in about $t = \boxed{1\frac{1}{5}} = \boxed{1.2}$ hours if they work together. Since there are 60 minutes in one hour and $(1.2)(60) = 72$, the answer could alternatively be expressed as 72 minutes. Check the answer: To find the fraction of the work that each person does, divide 1.2 hours by the time it would take each person to work alone. For Pete, the fraction of the work done is $1.2 \div 3$, and for Brenda the fraction of the work done is $1.2 \div 2$. If you add these fractions together, the sum must add up to one (because all of the work gets done): $1.2 \div 3 + 1.2 \div 2 = 0.4 + 0.6 = 1$. It should also make sense that the work gets done in less time (1.2 hours) when they work together than it would if either worked alone (3 hours or 2 hours).

PROBLEM #15

If the system below is in static equilibrium, what is the value of w_2?

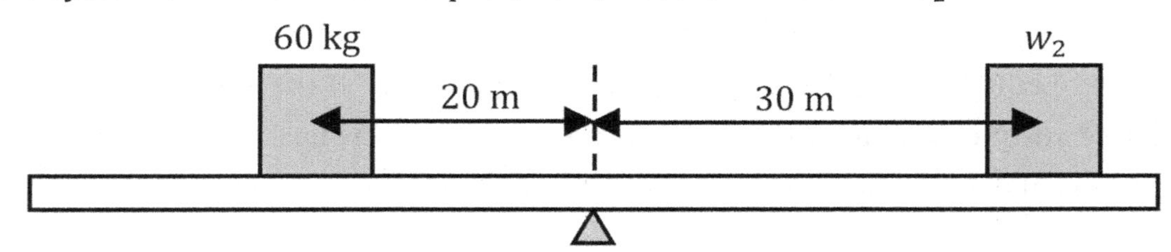

SOLUTION #15

w_2 = the weight of the box on the right side of the fulcrum

When a lever is in static equilibrium, the counterclockwise torques and clockwise torques are equal. (Note that the weights are **not** equal.) Set the sum of the counterclockwise torques equal to the sum of the clockwise torques, where each torque equals weight times lever arm. (See the side note below regarding weight.)

$$w_1 r_1 = w_2 r_2$$
$$(60)(20) = (w_2)(30)$$
$$1200 = 30w_2$$
$$\frac{1200}{30} = 40 = w_2$$

The box on the right side of the fulcrum has a "weight" of $w_2 = \boxed{40}$ kg. Check the answer: The counterclockwise torque is $(60)(20) = 1200$ and the clockwise torque is $(40)(30) = 1200$.

Side note: Strictly speaking, the weight (in Newtons) equals the mass (in kilograms) times gravity (9.81 m/s^2). However, since every term in the equation includes weight, gravity cancels out. This means that we may choose to work with mass instead of weight (so there is no reason to multiply all of the masses by 9.81 m/s^2). The answer given of 40 kg is technically the "mass" of the box on the right, and the weight of the box on the right is technically $w_2 = (40)(9.81) = 392.4$ N. However, it's not uncommon for people to use the words "mass" and "weight" interchangeably. In this book, we will focus on the algebra. It isn't necessary to have any prior knowledge of science in order to solve the problems in this book. (However, in a science course you better be careful to distinguish between mass and weight.)

PROBLEM #16

A rectangle has a perimeter of 21 in. The width of the rectangle is twice as long as the height of the rectangle. What are the width and height of the rectangle?

SOLUTION #16

$$h = \text{the height of the rectangle in inches}$$

The width of the rectangle is twice as long as the height of the rectangle. It follows that:

$$w = 2h = \text{the width of the rectangle in inches}$$

Plug the given perimeter into the formula for the perimeter of a rectangle.

$$P = 2w + 2h$$

$$21 = 2w + 2h$$

Substitute the equation $w = 2h$ into the equation for the perimeter.

$$21 = 2(2h) + 2h$$

$$21 = 4h + 2h$$

$$21 = 6h$$

$$\frac{21}{6} = \frac{7}{2} = 3.5 = h$$

The height of the rectangle is $h = \boxed{3.5}$ in. The width of the rectangle is $w = 2h = 2(3.5) = \boxed{7}$ in. Check the answers: The perimeter is $P = 2w + 2h = 2(7) + 2(3.5) = 14 + 7 = 21$ in.

PROBLEM #17

One number is four times another number. The numbers have a product of 324. If both numbers are positive, what are the numbers?

SOLUTION #17

$$x = \text{the smaller number}$$

The larger number is four times the smaller number. It follows that:

$$4x = \text{the larger number}$$

The product equals 324. Multiply the two numbers together.

$$(x)(4x) = 324$$

Recall that a variable times itself equals the variable squared: $xx = x^2$.

$$4x^2 = 324$$

Divide both sides of the equation by 4.

$$x^2 = \frac{324}{4}$$

$$x^2 = 81$$

Squareroot both sides of the equation.

$$\sqrt{x^2} = \sqrt{81}$$

Recall that $\sqrt{x^2} = x$.

$$x = 9$$

The smaller number is $x = 9$ and the larger number equals $4x = 4(9) = 36$. The two numbers are $\boxed{9}$ and $\boxed{36}$. Check the answers: The product is $(9)(36) = 324$ and 9 times 4 equals 36.

PROBLEM #18

A girl writes down three consecutive odd numbers. The sum of the first two numbers is greater than the third number by 75. What are the numbers?

SOLUTION #18

$$x = \text{the smallest number}$$

Consecutive odd numbers have a difference of 2. Since the smallest number is x, it follows that:

$$x + 2 = \text{the second number}$$

$$x + 4 = \text{the largest number}$$

The sum of the first two numbers is greater than the third number by 75. Add up the first two numbers and set that equal to the third number plus 75. (Alternatively, you could subtract 75 from the sum and set that expression equal to the third number.)

$$x + (x + 2) = (x + 4) + 75$$

Isolate the unknown. Combine like terms.

$$x + x + 2 = x + 4 + 75$$

$$2x + 2 = x + 79$$

$$2x - x = 79 - 2$$

$$x = 77$$

The first number is $x = \boxed{77}$, the second number is $x + 2 = 77 + 2 = \boxed{79}$, and the last number is $x + 4 = 77 + 4 = \boxed{81}$. Check the answers: 77, 79, and 81 are consecutive odd numbers. The sum of the first two numbers is $77 + 79 = 156$, and 156 is larger than 81 by 75 since $156 - 75 = 81$ (or $156 = 81 + 75$).

PROBLEM #19

Sheila is 15 years older than Jackie. Five years ago, Sheila was four times as old as Jackie. What are their ages now?

SOLUTION #19

Jackie is younger than Sheila.

$$x = \text{Jackie's age now}$$

Sheila is 15 years older than Jackie.

$$x + 15 = \text{Sheila's age now}$$

The second sentence refers to their ages 5 year ago. Subtract 5 from their current ages to find their ages 5 years ago.

$$x - 5 = \text{Jackie's age 5 years ago}$$

$$(x + 15) - 5 = x + 10 = \text{Sheila's age 5 years ago}$$

5 years ago, Sheila was four times as old as Jackie. Multiply Jackie's age by 4, but use their ages 5 years ago.

$$(x + 10) = 4(x - 5)$$

Distribute the 4.

$$x + 10 = 4x - 20$$

Combine like terms.

$$10 + 20 = 4x - x$$

$$30 = 3x$$

$$\frac{30}{3} = 10 = x$$

Jackie is $x = \boxed{10}$ years old now and Sheila is $x + 15 = 10 + 15 = \boxed{25}$ years old now. Check the answers: Presently, Sheila (25) is 15 years older than Jackie (10). Five years ago, Sheila was $25 - 5 = 20$ years old and Jackie was $10 - 5 = 5$ years old. Five years ago, Sheila (20) was four times as old as Jackie (5).

PROBLEM #20

The sum of the digits of a three-digit number is 16. The tens digit is two more than the units digit, and the units digit is triple the hundreds digit. What is the number?

SOLUTION #20

The hundreds digit must be smaller than the units digit and the tens digit.

$$x = \text{the hundreds digit of the number}$$

The units digit is triple the hundreds digit. Since the hundreds digit is x, it follows that:

$$3x = \text{the units digit of the number}$$

The tens digit is two more than the units digit. Since the units digit is $3x$, it follows that:

$$3x + 2 = \text{the tens digit of the number}$$

The sum of the digits is 16. Add the three digits together.

$$x + 3x + (3x + 2) = 16$$

$$7x + 2 = 16$$

$$7x = 14$$

$$x = \frac{14}{7} = 2$$

The hundreds digit is $x = 2$, the tens digit is $3x + 2 = 3(2) + 2 = 6 + 2 = 8$, and the units digit is $3x = 3(2) = 6$. The number is $100(2) + 10(8) + 1(6) = 200 + 80 + 6 = \boxed{286}$. Check the answer: The digits add up to $2 + 8 + 6 = 16$, the tens digit is two more than the units digit (since $8 - 6 = 2$), and the units digit is triple the hundreds digit (since 6 equals 3 times 2).

PROBLEM #21

1200 female students attend a college where the ratio of female students to male students is 8:5. How many students attend the college?

SOLUTION #21

$$x = \text{the number of male students attending the college}$$

This problem tells us the number of female students.

$$1200 = \text{the number of female students attending the college}$$

The ratio of female students to male students is 8:5. This ratio can be expressed as the fraction $\frac{8}{5}$. Multiply the number of male students by this fraction to get the number of female students.

$$\frac{8x}{5} = 1200$$

Multiply both sides of the equation by 5.

$$8x = (5)(1200)$$

$$8x = 6000$$

$$x = \frac{6000}{8} = 750$$

The question asked for the total number of students attending the college. Add the number of male students (750) to the number of female students (1200) to get the total number of students.

$$x + 1200 = 750 + 1200 = 1950$$

There are $\boxed{1950}$ students attending the college (1200 female students and 750 male students). Check the answer: The ratio of female students to male students is $\frac{1200}{750} = \frac{1200 \div 150}{750 \div 150} = \frac{8}{5}$.

PROBLEM #22

A wallet contains twenty-dollar bills, ten-dollar bills, five-dollar bills, and one-dollar bills. There are nine times as many one-dollar bills as ten-dollar bills, there are two more twenty-dollar bills than ten-dollar bills, and the value of the five-dollar bills is twice the value of the twenty-dollar bills. The total value of the money in the wallet is $357. How many bills of each kind are in the wallet?

SOLUTION #22

There are fewer ten-dollar ($10) bills than twenty-dollar ($20) bills, five-dollar ($5) bills, or one-dollar ($1) bills.

$$x = \text{the number of ten-dollar (\$10) bills}$$

There are nine times as many one-dollar ($1) bills as ten-dollar ($10) bills.

$$9x = \text{the number of one-dollar (\$1) bills}$$

There are two more twenty-dollar ($20) bills than ten-dollar ($10) bills.

$$x + 2 = \text{the number of twenty-dollar (\$20) bills}$$

Since one ten-dollar bill is worth $10, multiply the number of ten-dollar bills by 10. Since a one-dollar bill is worth $1, multiply the number of one-dollar bills by 1. Since one twenty-dollar bills is worth $20, multiply the number of twenty-dollar bills by 20.

$$10x = \text{the value of the ten-dollar (\$10) bills}$$

$$1(9x) = 9x = \text{the value of the one-dollar (\$1) bills}$$

$$20(x + 2) = 20x + 40 = \text{the value of the twenty-dollar (\$20) bills}$$

The value of the five-dollar ($5) bills is twice the value of the twenty-dollar ($20) bills. It would be incorrect to multiply the number of twenty-dollar ($20) bills by two: We must work with the value of the bills, not the number of bills.

$$2(20x + 40) = 40x + 80 = \text{the value of the five-dollar (\$5) bills}$$

The total value of the money in the wallet is $357. Add up the values of the bills.

$$10x + 9x + (20x + 40) + (40x + 80) = 357$$

$$79x + 120 = 357$$

$$79x = 237$$

$$x = \frac{237}{79} = 3$$

There are $9x = 9(3) = \boxed{27}$ one-dollar ($1) bills, $\frac{40x+80}{5} = \frac{40(3)+80}{5} = \frac{200}{5} = \boxed{40}$ five-dollar ($5) bills, $x = \boxed{3}$ ten-dollar ($10) bills, and $x + 2 = 3 + 2 = \boxed{5}$ twenty-dollar ($20) bills. Note: We divided the value of the five-dollar ($5) bills by 5 to get the number of five-dollar bills. Check the answers: The total value of the bills comes to $27(\$1) + 40(\$5) + 3(\$10) + 5(\$20) = \$27 + \$200 + \$30 + \$100 = \$357$.

PROBLEM #23

A saleslady sold necklaces for $75 each and bracelets for $25 each in a state where the sales tax was 10%. The saleslady sold 30 more bracelets than necklaces. The total amount paid was $2145. How many necklaces and how many bracelets did the saleslady sell?

SOLUTION #23

Fewer necklaces were sold than bracelets.

$$x = \text{the number of necklaces sold}$$

30 more bracelets were sold than necklaces.

$$x + 30 = \text{the number of bracelets sold}$$

To find the amount paid for the necklaces in dollars, multiply the number of necklaces (x) by 75. To find the amount paid for the bracelets in dollars, multiply the number of bracelets ($x + 30$) by 25.

$$75(x) = 75x = \text{the amount paid for the necklaces (before tax)}$$

$$25(x + 30) = 25x + 750 = \text{the amount paid for the bracelets (before tax)}$$

The subtotal equals total amount paid before tax.

$$75x + (25x + 750) = 100x + 750 = \text{the subtotal (before tax)}$$

A 10% tax was paid. Multiply the subtotal by 1.1 (for 110%). Why? The total amount paid was 100% + 10% = 110% because the tax adds 10% to the total cost. The total amount paid was \$2145.

$$(100x + 750)1.1 = 2145$$

$$110x + 825 = 2145$$

$$110x = 1320$$

$$x = \frac{1320}{110} = 12$$

The saleslady sold $x = \boxed{12}$ necklaces and $x + 30 = 12 + 30 = \boxed{42}$ bracelets. Check the answers: The subtotal is $\$75(12) + \$25(42) = \$900 + \$1050 = \$1950$, the 10% tax comes to $(0.1)(\$1950) = \$195$, and the total with tax is $\$1950 + \$195 = \$2145$.

PROBLEM #24

An investor purchased two different stocks. He paid a total of $4000. One stock made a profit of 15%, while the other stock suffered a loss of 6%. The net gain from the stocks was $243. How much money was invested in each stock?

SOLUTION #24

$x =$ the amount invested in the stock that made a profit of 15%

The total amount invested was \$4000. Part of this was invested in the stock that made a profit of 15% and the remainder was invested in the stock that suffered a loss of 6%. The two investments add up to \$4000. It follows that:

$4000 - x =$ the amount invested in the stock that suffered a loss of 6%

The interest earned (I) equals the principal invested (P) times the interest rate (r): $I = Pr$. Apply this formula to each investment. Divide by 100% to convert 15% and 6% into decimals: $\frac{15\%}{100\%} = 0.15$ and $\frac{6\%}{100\%} = 0.06$.

$0.15x =$ the profit from the stock that gained 15%

$0.06(4000 - x) = 240 - 0.06x =$ the loss from the stock that lost 6%

The total return was \$243. The profit minus the loss equals \$243. (If they were both profits, we would add them, but one was a loss, so we subtract the loss. You might think of the loss as a negative profit.)

$$0.15x - (240 - 0.06x) = 243$$

Distribute the minus sign. Note that $-1(-0.06x) = +0.06x$. The minus signs cancel.

$$0.15x - 240 + 0.06x = 243$$

Multiply by 100 to remove all of the decimals.

$$15x - 24{,}000 + 6x = 24{,}300$$

$$21x = 48{,}300$$

$$x = \frac{48{,}300}{21} = 2300$$

The investor purchased the stock that returned a 15% profit for $x = \boxed{2300}$ dollars and purchased the stock that suffered a 6% loss for $4000 - 2300 = \boxed{1700}$ dollars. Check the answers: The profit from the first stock is $(0.15)(\$2300) = \$345$. The loss from the second stock is $(0.06)(\$1700) = \$102$. The profit of \$345 minus the loss of \$102 yields a total return of $\$345 - \$102 = \$243$.

PROBLEM #25

Two numbers have an average value of 22 and a product of 475. What are the numbers?

SOLUTION #25

$$x = \text{the first number}$$

The numbers have a product of 475. It follows that:

$$\frac{475}{x} = \text{the second number}$$

The average value is 22. To find the average value of two numbers, add the numbers together and divide by two. (In general, divide by how many numbers you are adding together. In this case, we are adding two numbers, so we divide by two.)

$$\frac{x + \frac{475}{x}}{2} = 22$$

Multiply both sides of the equation by 2.

$$x + \frac{475}{x} = 44$$

Multiply both sides of the equation by x.

$$x^2 + 475 = 44x$$

This is a quadratic equation. Reorder the terms.

$$x^2 - 44x + 475 = 0$$

Compare this to the standard form, $ax^2 + bx + c = 0$, to see that $a = 1$, $b = -44$, and $c = 475$. Plug these values into the quadratic formula.

$$x = \frac{-b \pm \sqrt{b^2 - 4ac}}{2a} = \frac{-(-44) \pm \sqrt{(-44)^2 - 4(1)(475)}}{2(1)} = \frac{44 \pm \sqrt{1936 - 1900}}{2}$$

$$x = \frac{44 \pm \sqrt{36}}{2} = \frac{44 \pm 6}{2}$$

There are two possible answers corresponding to the plus and minus signs.

$$x = \frac{44 + 6}{2} = \frac{50}{2} = 25 \quad \text{or} \quad x = \frac{44 - 6}{2} = \frac{38}{2} = 19$$

The first number is $x = \boxed{19}$ and the second number is $\frac{475}{19} = \boxed{25}$. Check the answers: The average value is $\frac{19+25}{2} = \frac{44}{2} = 22$ and the product is $(19)(25) = 475$.

PROBLEM #26

A brother and sister are initially standing together. The brother grabs the sister's toy and begins running away with a speed of 2 m/s. Four seconds later, the sister realizes what happened and chases her brother with a speed of 3 m/s. Where does she catch her brother?

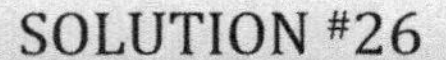

SOLUTION #26

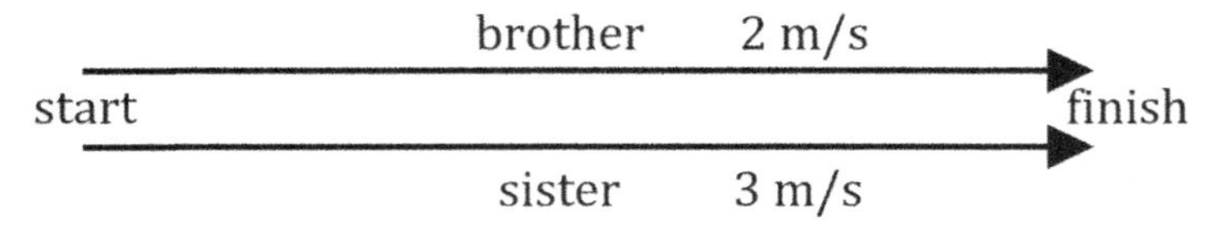

$t =$ the amount of time in seconds that has passed since the brother began running

$d =$ the distance that the brother has traveled when he is caught by his sister

The sister begins running four seconds after the brother began running. This means that the sister spends less time running.

$t - 4 =$ the amount of time that has passed since the sister began running

Distance (d) equals rate (r) times time (t): $d = rt$. The rates are 2 m/s and 3 m/s. Apply the rate equation to each person. Organize the information in a table.

	rate (m/s)	time (s)	distance (m)
brother	2	t	$2t$
sister	3	$t - 4$	$3(t - 4) = 3t - 12$

The brother and sister travel the same distance. Set their distances equal.

$$2t = 3t - 12$$
$$12 = 3t - 2t$$
$$12 = t$$

So far, we have found "when" the brother gets caught, but the question asked "where" the brother gets caught. The rate equation tells us how far the brother travels.

$$d = 2t = 2(12) = 24$$

The brother has traveled a distance of $d = \boxed{24}$ meters when his sister catches him. Check the answer: The distance traveled by the sister is $d = 3(t - 4) = 3(12 - 4) = 3(8) = 24$ meters, which agrees with the distance traveled by the brother. Since they both start in the same place and both finish in the same place, the distance must be the same for each.

PROBLEM #27

A solution of 60% hydrochloric acid is mixed with a solution of 35% hydrochloric acid. How much of each solution should be combined in order to make a 5-L mixture that is 50% hydrochloric acid?

SOLUTION #27

$x =$ the volume in liters of the solution that is 60% hydrochloric acid

The amount of hydrochloric acid (P) equals the decimal value of the percentage (c) times the volume of the solution (M): $P = cM$. Apply this formula to each solution. Divide the given percentages by 100% in order to convert them to decimals: $\frac{60\%}{100\%} = 0.6$, $\frac{35\%}{100\%} = 0.35$, and $\frac{50\%}{100\%}=0.5$. Organize the information in a table. The volume of the mixture is the sum of the volumes of the two solutions: $M_1 + M_2 = M_3 \rightarrow M_2 = 5 - x$.

	60% HCl	35% HCl	50% HCl
c (decimal)	0.6	0.35	0.5
M (liters)	x	$5 - x$	5
$P = cM$ (pure)	$0.6x$	$0.35(5 - x)$	$(0.5)(5) = 2.5$

The amount of pure stuff in the mixture is the sum of the amounts of the pure stuff in the two solutions: Add the values in the bottom row of the table.

$$0.6x + 0.35(5 - x) = 2.5$$

$$0.6x + 1.75 - 0.35x = 2.5$$

Multiply both sides of the equation by 100 in order to remove the decimal point.

$$60x + 175 - 35x = 250$$

$$25x + 175 = 250$$

$$25x = 75$$

$$x = \frac{75}{25} = 3$$

The volumes are $x = \boxed{3}$ liters of 60% hydrochloric acid and $5 - x = 5 - 3 = \boxed{2}$ liters of 35% hydrochloric acid. Check the answers: $P_1 + P_2 = 0.6(3) + 0.35(2) = 1.8 + 0.7 = 2.5 = P_3$ agrees with $P_3 = c_3M_3 = (0.5)(5) = 2.5$.

PROBLEM #28

Eileen could rake the leaves in a yard in 3 hours by herself. If Eileen and Yvonne work together, it takes them 2 hours to rake the leaves in the yard. How long would it take for Yvonne to rake the yard by herself?

SOLUTION #28

$t =$ the time it takes Yvonne to rake the yard by herself

Find the reciprocal of each given time in order to determine how much work each person could complete in one hour.

	Eileen	Yvonne	together
time to work (in hours)	3	t	2
fraction completed per hour	$\frac{1}{3}$	$\frac{1}{t}$	$\frac{1}{2}$

When people work together, add the reciprocals from the bottom row of the table.

$$\frac{1}{3} + \frac{1}{t} = \frac{1}{2}$$

The lowest common denominator of 3, t, and 2 is $6t$. Multiply both sides by $6t$.

$$\frac{6t}{3} + \frac{6t}{t} = \frac{6t}{2}$$

$$2t + 6 = 3t$$

$$6 = t$$

Yvonne could rake the leaves in the yard by herself in about $t = \boxed{6}$ hours. Check the answer: The fraction of the work that each person does equals 2 hours (the time it takes to work together) divided by the time it takes to work alone. Eileen does $\frac{2}{3}$ of the work and Yvonne does $\frac{2}{6}$ of the work. Since all of the work gets done, these fractions must add up to one: $\frac{2}{3} + \frac{2}{6} = \frac{2}{3} + \frac{1}{3} = 1$ (because $\frac{2}{6} = \frac{1}{3}$). It should also make sense that the work gets done in less time (2 hours) when they work together than it would if either worked alone (3 hours or 6 hours).

PROBLEM #29

If the system below is in static equilibrium, what is the value of r_3?

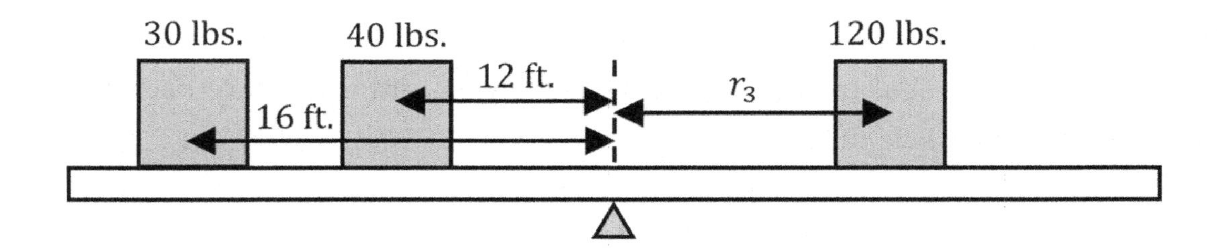

SOLUTION #29

r_3 = the distance from the fulcrum to the center of the 120-lb. box

When a lever is in static equilibrium, the counterclockwise torques and clockwise torques are equal. (Note that the weights are **not** equal.) Set the sum of the counter-clockwise torques equal to the sum of the clockwise torques, where each torque equals weight times lever arm.

$$w_1 r_1 + w_2 r_2 = w_3 r_3$$
$$(30)(16) + (40)(12) = (120)(r_3)$$
$$480 + 480 = 120 r_3$$
$$960 = 120 r_3$$
$$\frac{960}{120} = 8 = r_3$$

The center of the box on the right side of the fulcrum is $r_3 = \boxed{8}$ ft. from the fulcrum. Check the answer: The counterclockwise torque is $(30)(16) + (40)(12) = 480 + 480 = 960$ and the clockwise torque is $(120)(8) = 960$.

Note: Most math problems are intentionally not drawn to scale. Don't try to guess a distance from a picture. Illustrations label distances that are given. If a distance isn't given in a diagram, you need to solve for the distance using algebra (as we have done here).

PROBLEM #30

A circle has a circumference of 42 in. What is the area of the circle?

SOLUTION #30

$$r = \text{the radius of the circle in inches}$$

$$A = \text{the area of the circle in square inches}$$

Use the formula for the circumference of a circle to determine the radius.

$$C = 2\pi r$$

$$42 = 2\pi r$$

$$\frac{42}{2\pi} = \frac{21}{\pi} \approx \frac{21}{3.14159} \approx 6.68451 = r$$

Now use the formula for the area of a circle.

$$A = \pi r^2 = \pi(6.68451)^2 \approx 140$$

The area of the circle is $A \approx \boxed{140}$ in.^{2} Check the answer: Rewrite $C = 2\pi r$ as $r = \frac{C}{2\pi}$ and substitute this into $A = \pi r^2$ to get $A = \pi \left(\frac{C}{2\pi}\right)^2 = \pi \left(\frac{C^2}{4\pi^2}\right) = \frac{\pi C^2}{4\pi^2} = \frac{C^2}{4\pi} = \frac{(42)^2}{4\pi} = \frac{1764}{4\pi} = \frac{441}{\pi} \approx \frac{441}{3.14159} \approx 140$.

PROBLEM #31

Two positive numbers have a product of 275 and a difference of 14. What are the numbers?

SOLUTION #31

$$x = \text{the smaller number}$$

The difference is 14. This means that one number is larger than the other number by 14. Since the smaller number is x, it follows that:

$$x + 14 = \text{the larger number}$$

The product equals 275. When the numbers are multiplied together, they make 275.

$$x(x + 14) = 275$$

Distribute the x.

$$x^2 + 14x = 275$$

This is a quadratic equation. Subtract 275 from both sides in order to express this equation in standard form.

$$x^2 + 14x - 275 = 0$$

Compare this to the standard form, $ax^2 + bx + c = 0$, to see that $a = 1$, $b = 14$, and $c = -275$. Plug these values into the quadratic formula.

$$x = \frac{-b \pm \sqrt{b^2 - 4ac}}{2a} = \frac{-14 \pm \sqrt{14^2 - 4(1)(-275)}}{2(1)} = \frac{-14 \pm \sqrt{196 + 1100}}{2}$$

$$x = \frac{-14 \pm \sqrt{1296}}{2} = \frac{-14 \pm 36}{2}$$

There are two possible answers corresponding to the plus and minus signs.

$$x = \frac{-14 + 36}{2} = \frac{22}{2} = 11 \quad \text{or} \quad x = \frac{-14 - 36}{2} = \frac{-50}{2} = -25$$

We must discard -25 as a possible solution since the problem specifically stated that the numbers are "positive." (Although it turns out that 25 is one of the answers, -25 is not one of the answers.) If the first number is $x = 11$, then the second number is $x + 14 = 11 + 14 = 25$. Thus, we see that the two numbers are $\boxed{11}$ and $\boxed{25}$. Check the answers: 11 and 25 have a difference of $25 - 11 = 14$ and have a product of $(11)(25) = 275$.

PROBLEM #32

Two consecutive multiples of 12 have a sum of 420. What are the numbers?

SOLUTION #32

$$x = \text{the smaller number}$$

Multiples of 12 include 12, 24, 36, 48, and so on. Any two consecutive multiples of 12 will have a difference of 12. For example, if the numbers are 36 and 48, or if the numbers are 72 and 84, the difference is 12. Since the smaller number is x, it follows that:

$$x + 12 = \text{the larger number}$$

The sum of the numbers is 420. Add the numbers together.

$$x + (x + 12) = 420$$

Isolate the unknown. Combine like terms.

$$x + x + 12 = 420$$

$$2x + 12 = 420$$

$$2x = 408$$

$$x = \frac{408}{2} = 204$$

The smaller number is $x = 204$ and the larger number is $x + 12 = 204 + 12 = 216$. The two numbers are $\boxed{204}$ and $\boxed{216}$. Check the answers: 204 and 216 are multiples of 12 since $204 \div 12 = 17$ and $216 \div 12 = 18$ (and they are consecutive multiples because 204 is the 17th multiple and 216 is the 18th multiple). The sum of 204 and 216 equals $204 + 216 = 420$.

PROBLEM #33

Jamie is currently 25 years older than Alex. In four years, Jamie will be three times as old as Alex was five years ago. What are their ages now?

SOLUTION #33

Alex is younger than Jamie.

$$x = \text{Alex's age now}$$

Jamie is currently 25 years older than Alex. It follows that:

$$x + 25 = \text{Jamie's age now}$$

The problem is asking about Alex's age five years ago. Subtract 5 from Alex's age to find Alex's age five years ago.

$$x - 5 = \text{Alex's age 5 years ago}$$

The problem is also asking about Jamie's age in four years. Add 4 to Jamie's age to find Jamie's age in four years.

$$(x + 25) + 4 = x + 29 = \text{Jamie's age in 4 years}$$

Jamie's age in 4 years (which will be $x + 29$) will be 3 times how old Alex was 5 years ago (which was $x - 5$). The problem is saying that $(x + 29)$ is 3 times $(x - 5)$.

$$(x + 29) = 3(x - 5)$$

Distribute the 3.

$$x + 29 = 3x - 15$$

Combine like terms.

$$29 + 15 = 3x - x$$

$$44 = 2x$$

$$\frac{44}{2} = 22 = x$$

Alex is $x = \boxed{22}$ years old now and Jamie is $x + 25 = 22 + 25 = \boxed{47}$ years old now. Check the answers: Presently, Jamie (47) is 25 years older than Alex (22). In 4 years, Jamie will be $47 + 4 = 51$ years old. Five years ago, Alex was $22 - 5 = 17$ years old. The problem asked us to compare Jamie's age in 4 years (which will be 51) to Alex's age 5 years ago (which was 17): Observe that $51 = 3(17)$.

PROBLEM #34

The digits of a two-digit number differ by four. When the digits are reversed, the reversed number is 75% larger than the original number. What is the number?

SOLUTION #34

The tens digit must be smaller than the units digit because the number is larger when the digits are reversed.

$$x = \text{the tens digit of the original number}$$

The digits differ by 4. Since the units digit is larger than the tens digit, it follows that:

$$x + 4 = \text{the units digit of the original number}$$

To make the original number, multiply the tens digit by 10 and the units digit by 1.

$$10(x) + 1(x + 4) = 10x + x + 4 = 11x + 4 = \text{the original number}$$

To make the reversed number, swap the tens digit with the units digit.

$$x + 4 = \text{the tens digit of the reversed number}$$

$$x = \text{the units digit of the reversed number}$$

$$10(x + 4) + 1(x) = 10x + 40 + x = 11x + 40 = \text{the reversed number}$$

The reversed number is 75% larger than the original number. This means that the reversed number equals the original number times 175%. That's because the original number corresponds to 100%, and 75% more than that makes 175%. Divide 175% by 100% in order to convert it to a decimal: $\frac{175\%}{100\%} = 1.75$.

$$11x + 40 = 1.75(11x + 4)$$

$$11x + 40 = 19.25x + 7$$

Multiply both sides of the equation by 4 to remove the decimal point. (You could multiply by 100 instead if you wish, but you will need to work with larger numbers.)

$$44x + 160 = 77x + 28$$

$$160 - 28 = 77x - 44x$$

$$132 = 33x$$

$$\frac{132}{33} = 4 = x$$

The tens digit is $x = 4$ and the units digit equals $x + 4 = 4 + 4 = 8$. The original number is $10(4) + 8 = \boxed{48}$. Check the answer: The reversed number is $10(8) + 4 = 84$. The ratio $\frac{84}{48} = \frac{7}{4} = 1.75 = 175\%$ shows that 84 is 75% larger than 48.

PROBLEM #35

A bottle contains quarters, dimes, nickels, and pennies. The ratio of quarters to dimes is 8:3, the ratio of nickels to quarters is 9:5, and the ratio of pennies to quarters is 7:2. The total value of the coins in the bottle is $16.50. How many coins of each kind are in the bottle?

SOLUTION #35

$$x = \text{the number of dimes}$$

The ratio of quarters to dimes is 8:3. The ratio of nickels to quarters is 9:5. The ratio of pennies to quarters is 7:2. These ratios can be expressed as the fractions $\frac{8}{3}$, $\frac{9}{5}$, and $\frac{7}{2}$. Multiply the coin in the denominator by this fraction to get the number of coins in the numerator. For example, in "quarters to dimes," the dimes are in the denominator.

$$\frac{8x}{3} = \text{the number of quarters}$$

$$\frac{9}{5}\left(\frac{8x}{3}\right) = \frac{72x}{15} = \frac{72x \div 3}{15 \div 3} = \frac{24x}{5} = \text{the number of nickels}$$

$$\frac{7}{2}\left(\frac{8x}{3}\right) = \frac{56x}{6} = \frac{56x \div 2}{6 \div 2} = \frac{28x}{3} = \text{the number of pennies}$$

Multiply by the value of each coin. For example, the value of a quarter is 25 cents.

$$25\left(\frac{8x}{3}\right) = \frac{200x}{3} = \text{the value of the quarters}$$

$$10(x) = 10x = \text{the value of the dimes}$$

$$5\left(\frac{24x}{5}\right) = 24x = \text{the value of the nickels}$$

$$1\left(\frac{28x}{3}\right) = \frac{28x}{3} = \text{the value of the pennies}$$

The total value of the coins in the bottle is \$16.50. Multiply \$16.50 by 100 to convert this to 1650 cents. Add up the values of the coins and set this equal to 1650 cents.

$$\frac{200x}{3} + 10x + 24x + \frac{28x}{3} = 1650 \quad \text{(multiply both sides by 3)}$$

$$200x + 30x + 72x + 28x = 4950$$

$$330x = 4950$$

$$x = \frac{4950}{330} = 15$$

There are $\frac{8x}{3} = \frac{8(15)}{3} = \boxed{40}$ quarters, $x = \boxed{15}$ dimes, $\frac{24x}{5} = \frac{24(15)}{5} = \boxed{72}$ nickels, and $\frac{28x}{3} = \frac{28(15)}{3} = \boxed{140}$ pennies. Check the answers: The total value of the coins in the bottle is $25(40) + 10(15) + 5(72) + 1(140) = 1650$.

PROBLEM #36

A website sold t-shirts. Some of the t-shirts sold at a regular price of $15, while other t-shirts were sold at a 20% discount. Two-thirds of the purchases were made using the 20% discount. No sales tax was charged. The total amount paid was $7020. How many t-shirts were sold at regular price and how many were sold at a discount?

SOLUTION #36

Fewer t-shirts were sold at regular price than were sold at a discount.

$x =$ the number of t-shirts sold at the regular price

Two-thirds of the t-shirts were sold at a 20% discount. This means that one-third of the t-shirts were sold at regular price. Since one-third were sold at regular price and two-thirds were sold at a discount, it should be clear that twice as many were sold at a discount than were sold at regular price.

$2x =$ the number of t-shirts sold at a 20% discount

To find the amount paid for the t-shirts that were sold at regular price, multiply the number of t-shirts sold at regular price (x) by 15 (since the regular price is \$15).

$15x =$ the amount paid for t-shirts purchased at the regular price

The other t-shirts were purchased at a 20% discount. Divide 20% by 100% in order to convert it to a decimal: $\frac{20\%}{100\%} = 0.2$. Multiply \$15 by 0.2 to determine the discount: $(0.2)(\$15) = \$3$. Subtract the discount (\$3) from the regular price (\$15) to determine the sale price: $\$15 - \$3 = \$12$. To find the amount paid for the t-shirts that were sold at a discount, multiply the number of t-shirts sold at the discount ($2x$) by 12.

$12(2x) = 24x =$ the amount paid for t-shirts purchased at a 20% discount

The total amount paid was \$7020. Add the amounts paid together.

$$15x + 24x = 7020$$

$$39x = 7020$$

$$x = \frac{7020}{39} = 180$$

The website sold $x = \boxed{180}$ t-shirts at the regular price of \$15 each and $2x = 2(180) = \boxed{360}$ t-shirts at a sale price of \$12 (which is a 20% discount). Check the answers: The total amount paid is $\$15(180) + \$12(360) = \$2700 + \$4320 = \$7020$. The total number of t-shirts sold is $180 + 360 = 540$. Two-thirds of the total number of t-shirts sold equals $\frac{2}{3}(540) = 360$, which shows that two-thirds of the t-shirts sold were sold at a 20% discount.

PROBLEM #37

A businessman invested $6400 in a stock and $3200 in a savings account. The interest rate earned by the stock was triple the interest rate earned by the savings account. The total return was $1120. What interest rate was earned from the stock and what interest rate was earned from the savings account?

SOLUTION #37

The interest rate was smaller for the savings account than for the stock.

$x =$ the interest rate (in decimal form) earned from the savings account

The stock's interest rate was triple the savings account's interest rate.

$3x =$ the interest rate (in decimal form) earned from the stock

The interest earned (I) equals the principal invested (P) times the interest rate (r): $I = Pr$. Apply this formula to each investment. The principal invested in the stock was \$6400 and the principal invested in the savings account was \$3200.

$6400(3x) = 19{,}200x =$ the interest earned from the stock

$3200x =$ the interest earned from the savings account

The total return was \$1120. Add the two returns together.

$$19{,}200x + 3200x = 1120$$

$$22{,}400x = 1120$$

$$x = \frac{1120}{22{,}400} = 0.05$$

The interest earned from the savings account was $x = 0.05 = \boxed{5\%}$ and the interest earned from the stock was $3x = 3(0.05) = 0.15 = \boxed{15\%}$. (Multiply by 100% in order to convert from a decimal to a percentage.) Check the answers: The total return is $0.05(\$3200) + 0.15(\$6400) = \$160 + \$960 = \$1120$.

PROBLEM #38

A girl is traveling to her grandma's house, which is 180 miles away. The girl wishes to reach her grandma's house in 4 hours. The girl spends the first 3 hours of her trip traveling 40 mph. How fast does the girl need to drive for the remainder of the trip in order to arrive on time?

SOLUTION #38

$r_2 =$ the rate for the second part of the trip

$d_1 =$ the distance for the first part of the trip

$d_2 =$ the distance for the second part of the trip

Distance (d) equals rate (r) times time (t): $d = rt$. We will apply the rate equation to each part of the trip. For the first part of the trip, the time is $t_1 = 3$ hours and the rate is $r_1 = 40$ mph. The distance traveled for the first part of the tip is:

$$d_1 = r_1 t_1 = (40)(3) = 120$$

The total distance traveled (for both trips combined) is $d_{total} = 180$ miles, and the total time for the trip (for both trips combined) is $t_{total} = 4$ hours. Therefore, the distance and time for the second part of the trip are:

$$d_2 = d_{total} - d_1 = 180 - 120 = 60$$

$$t_2 = t_{total} - t_1 = 4 - 3 = 1$$

Now apply the rate equation to the second part of the trip.

$$d_2 = r_2 t_2$$

$$60 = r_2(1)$$

$$60 = r_2$$

In order to reach her grandma's house on time, the girl needs to travel $r_2 = \boxed{60}$ mph for the second part of the trip. Check the answer: The distances are $d_1 = r_1 t_1 = (40)(3) = 120$, $d_2 = r_2 t_2 = (60)(1) = 60$, and $d_{total} = d_1 + d_2 = 120 + 60 = 180$. The times are $t_1 = 3$, $t_2 = 1$, and $t_{total} = t_1 + t_2 = 3 + 1 = 4$.

PROBLEM #39

Approximately, how much water must be added to 360 mL of a solution of 60% nitric acid in order to dilute the solution down to 40%?

SOLUTION #39

$x =$ the volume of water in milliliters that needs to be added

The amount of pure nitric acid (P) equals the decimal value of the percentage (c) times the volume of the mixture (M): $P = cM$. Apply this formula to each mixture. Divide the given percentages by 100% in order to convert them to decimals: $\frac{60\%}{100\%} = 0.6$ and $\frac{40\%}{100\%} = 0.4$. Organize the information in a table. The volume of the mixture is the sum of the volumes of the two solutions: $M_1 + M_2 = M_3$. Water is 0% nitric acid.

	60% nitric acid	water (0% nitric acid)	40% nitric acid
c (decimal)	0.6	0	0.4
M (mL)	360	x	$360 + x$
$P = cM$ (pure)	$(0.6)(360) = 216$	$0x = 0$	$0.4(360 + x)$

The amount of pure stuff in the mixture is the sum of the amounts of the pure stuff in the two solutions: Add the values in the bottom row of the table.

$$216 + 0 = 0.4(360 + x)$$
$$216 = 144 + 0.4x$$
$$72 = 0.4x$$

If you wish to remove the decimal, multiply both sides of the equation by 10.

$$720 = 4x$$
$$\frac{720}{4} = 180 = x$$

The volume of water needed is $x = \boxed{180}$ mL. Check the answer: $P_1 + P_2 = 216 + 0 = 216$ agrees with $P_3 = c_3M_3 = 0.4(360 + x) = 0.4(360 + 180) = 0.4(540) = 216$.

Side note: Why does the problem say "approximately"? It has to do with chemistry. When two solutions are mixed together, the volume isn't *exactly* equal to the sum of the volumes. The equation $M_1 + M_2 = M_3$ is approximately true, but not exact.

PROBLEM #40

Warren could wash a truck in 33 min. Lisa could wash the same truck in 22 min. Sam could wash the same truck in 11 min. If they work together to wash the truck, about how long would it take?

SOLUTION #40

$t =$ the time it takes Warren, Lisa, and Sam to wash the truck together

Find the reciprocal of each given time in order to determine how much work each person could complete in one minute.

	Warren	Lisa	Sam	together
time to work (in minutes)	33	22	11	t
fraction completed per minute	$\frac{1}{33}$	$\frac{1}{22}$	$\frac{1}{11}$	$\frac{1}{t}$

When people work together, add the reciprocals from the bottom row of the table.

$$\frac{1}{33} + \frac{1}{22} + \frac{1}{11} = \frac{1}{t}$$

The lowest common denominator of 33, 22, 11, and t is $66t$. Multiply both sides by $66t$.

$$\frac{66t}{33} + \frac{66t}{22} + \frac{66t}{11} = \frac{66t}{t}$$

$$2t + 3t + 6t = 66$$

$$11t = 66$$

$$t = \frac{66}{11} = 6$$

Warren, Lisa, and Sam could wash the truck in about $t = \boxed{6}$ minutes if they work together. Check the answer: The fraction of the work that each person does equals 6 minutes (the time it takes to work together) divided by the time it takes to work alone. Warren does $\frac{6}{33}$ of the work, Lisa does $\frac{6}{22}$ of the work, and Sam does $\frac{6}{11}$ of the work. Since all of the work gets done, these fractions must add up to one: $\frac{6}{33} + \frac{6}{22} + \frac{6}{11} = \frac{12}{66} + \frac{18}{66} + \frac{36}{66} = \frac{66}{66} = 1$. It should also make sense that the work gets done in less time (6 min.) when they work together than it would if either worked alone (33 min., 22 min., or 11 min.).

PROBLEM #41

If the system below is in static equilibrium, what is the value of w_2?

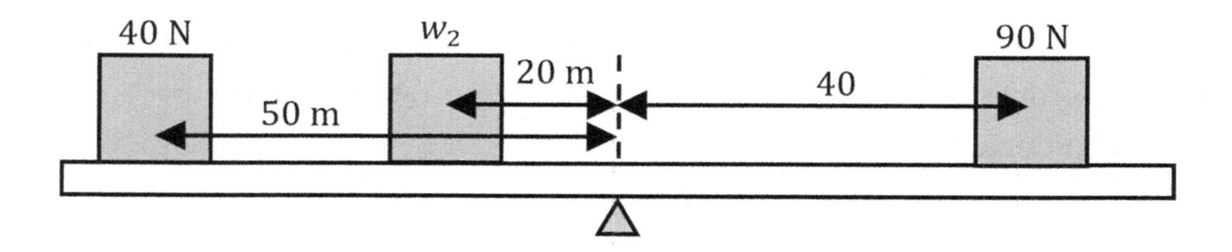

SOLUTION #41

$w_2 =$ the weight of the second box on the left side of the fulcrum

When a lever is in static equilibrium, the counterclockwise torques and clockwise torques are equal. (Note that the weights are **not** equal.) Set the sum of the counterclockwise torques equal to the sum of the clockwise torques, where each torque equals weight times lever arm.

$$w_1 r_1 + w_2 r_2 = w_3 r_3$$
$$(40)(50) + (w_2)(20) = (90)(40)$$
$$2000 + 20w_2 = 3600$$
$$20w_2 = 1600$$
$$w_2 = \frac{1600}{20} = 80$$

The second box on the left side of the fulcrum has a weight of $w_2 = \boxed{80}$ Newtons. Check the answer: The counterclockwise torque is $(40)(50) + (80)(20) = 2000 + 1600 = 3600$ and the clockwise torque is $(90)(40) = 3600$.

PROBLEM #42

The base of a triangle is 4 cm longer than its height. The area of the triangle is 48 cm^2. What are the base and height of the triangle?

SOLUTION #42

$b =$ the base of the triangle in centimeters

$h =$ the height of the triangle in centimeters

Plug the given area into the formula for the area of a triangle, which is $A = \frac{1}{2}bh$.

$$48 = \frac{1}{2}bh \quad \text{(area equation)}$$

The base is 4 cm longer than the height. Add 4 cm to the height to make the base.

$$b = h + 4 \quad \text{(base equation)}$$

Substitute this expression for the base into the area equation.

$$48 = \frac{1}{2}(h + 4)h$$

Multiply both sides of the equation by 2.

$$96 = (h + 4)h$$

Distribute.

$$96 = h^2 + 4h$$

This is a quadratic equation. Reorder the terms in order to express this equation in standard form.

$$-h^2 - 4h + 96 = 0$$

Compare this to the standard form, $ax^2 + bx + c = 0$, to see that $a = -1$, $b = -4$, and $c = 96$. Plug these values into the quadratic formula.

$$x = \frac{-b \pm \sqrt{b^2 - 4ac}}{2a} = \frac{-(-4) \pm \sqrt{(-4)^2 - 4(-1)(96)}}{2(-1)} = \frac{4 \pm \sqrt{16 + 384}}{-2}$$

$$x = \frac{4 \pm \sqrt{400}}{-2} = \frac{4 \pm 20}{-2}$$

There are two possible answers corresponding to the plus and minus signs.

$$x = \frac{4 + 20}{-2} = \frac{24}{-2} = -12 \text{ (not possible)} \quad \text{or} \quad x = \frac{4 - 20}{-2} = \frac{-16}{-2} = 8 \text{ (possible)}$$

The height of the triangle is $h = \boxed{8}$ cm and the base is $b = h + 4 = 8 + 4 = \boxed{12}$ cm.

Check the answers: The area is $A = \frac{1}{2}bh = \frac{1}{2}(12)(8) = 48$ cm.^{2}

PROBLEM #43

Violet had four times as many grapes as Zack. After Violet gave 27 grapes to Zack, they had the same number of grapes. How many grapes did they each have to begin with?

SOLUTION #43

In the beginning, Zack had fewer grapes than Violet.

$$x = \text{the number of grapes that Zack had to begin with}$$

In the beginning, Violet had four times as many grapes as Zack. It follows that:

$$4x = \text{the number of grapes that Violet had to begin with}$$

Violet gave 27 grapes to Zack. Add 27 grapes to Zack's number and subtract 27 from Violet's number to determine how many grapes each has now.

$$x + 27 = \text{the number of grapes that Zack has now}$$

$$4x - 27 = \text{the number of grapes that Violet has now}$$

Now that Violet has given 27 grapes to Zack, they have the same number of grapes. Set the current number of grapes for Zack and Violet equal to each other.

$$x + 27 = 4x - 27$$

Combine like terms. Subtract x from both sides and add 27 to both sides. Note that 27 doesn't cancel: Instead, you get $27 + 27 = 54$.

$$27 + 27 = 4x - x$$

$$54 = 3x$$

$$\frac{54}{3} = 18 = x$$

Zack had $x = \boxed{18}$ grapes to begin with and Violet had $4x = 4(18) = \boxed{72}$ grapes to begin with. Check the answers: After Violet gave 27 grapes to Zack, Violet had $72 - 27 = 45$ grapes and Zack had $18 + 27 = 45$ grapes.

PROBLEM #44

Amy wrote down two consecutive odd numbers on a sheet of paper. Amy squared each of the numbers. When she subtracted the squared numbers, she got 88. What are the numbers?

SOLUTION #44

$$x = \text{the smaller number}$$

Consecutive odd numbers have a difference of 2. Since the smaller number is x, it follows that:

$$x + 2 = \text{the larger number}$$

Amy squared each number.

$$x^2 = \text{the square of the smaller number}$$
$$(x + 2)^2 = \text{the square of the larger number}$$

Multiply $(x + 2)$ by itself. Apply the f.o.i.l. method.

$$(x + 2)^2 = (x + 2)(x + 2) = x^2 + 2x + 2x + 4 = x^2 + 4x + 4$$

Therefore, we may write:

$$(x + 2)^2 = x^2 + 4x + 4 = \text{the square of the larger number}$$

When Amy subtracted the squared numbers, she got 88.

$$(x^2 + 4x + 4) - x^2 = 88$$
$$x^2 + 4x + 4 - x^2 = 88$$
$$4x + 4 = 88$$
$$4x = 84$$
$$x = \frac{84}{4} = 21$$

The smaller number is $x = 21$ and the larger number is $x + 2 = 21 + 2 = 23$. The two numbers are $\boxed{21}$ and $\boxed{23}$. Check the answers: $23^2 - 21^2 = (23)(23) - (21)(21) = 529 - 441 = 88$.

PROBLEM #45

Theresa is 23 years old. Her grandmother is 71 years old. How many years ago was her grandmother five times as old as Theresa?

SOLUTION #45

$x =$ the number of years ago that her grandmother was 5 times as old as Theresa

Their current ages are:

$$23 = \text{Theresa's age now}$$

$$71 = \text{her grandmother's age now}$$

Subtract x from their current ages to find their ages x years ago.

$$23 - x = \text{Theresa's age } x \text{ years ago}$$

$$71 - x = \text{her grandmother's age } x \text{ years ago}$$

Set her grandmother's age equal to 5 times Theresa's age, but use their ages x years ago.

$$5(23 - x) = 71 - x$$

Distribute the 5.

$$115 - 5x = 71 - x$$

Combine like terms.

$$115 - 71 = -x + 5x$$

$$44 = 4x$$

$$\frac{44}{4} = 11 = x$$

$x = \boxed{11}$ years ago, her grandmother was five times as old as Theresa. Check the answer: Presently, Theresa is 23 years old and her grandmother is 71 years old. Going back 11 years, Theresa was $23 - 11 = 12$ years old and her grandmother was $71 - 11 = 60$ years old. 11 years ago, her grandmother was $\frac{60}{12} = 5$ times as old as Theresa.

PROBLEM #46

A number has two digits. When the digits are reversed, the reversed number is larger than the original number by 27. The reversed number and the original number have a sum of 99. What is the number?

SOLUTION #46

It is convenient to work with two variables. We will need two equations. (The last four examples of Chapter 3 and the last page of Chapter 2 involve two variables.)

$$x = \text{the tens digit of the original number}$$

$$y = \text{the units digit of the original number}$$

To make the original number, multiply the tens digit by 10 and the units digit by 1.

$$10x + y = \text{the original number}$$

To make the reversed number, swap the tens digit with the units digit.

$$10y + x = \text{the reversed number}$$

The reversed number is larger than the original number by 27.

$$(10y + x) = (10x + y) + 27$$

$$9y - 9x = 27 \quad \text{(first equation)}$$

The reversed number and the original number have a sum of 99.

$$(10y + x) + (10x + y) = 99$$

$$11y + 11x = 99 \quad \text{(second equation)}$$

Divide the first equation by 9 and the second equation by 11. See the last page of Chapter 2 (but note that for this problem it turns out to be simpler to divide).

$$9y - 9x = 27 \quad \to \quad y - x = 3$$

$$11y + 11x = 99 \quad \to \quad y + x = 9$$

Add the equations together. The $-x$ and $+x$ terms cancel out in the addition.

$$y - x + y + x = 12$$

$$2y = 12 \quad \to \quad y = \frac{12}{2} = 6$$

$$y + x = 9 \quad \to \quad 6 + x = 9 \quad \to \quad x = 9 - 6 = 3$$

The tens digit is $x = 3$ and the units digit is $y = 6$. The original number is $10(3) + 6 = 30 + 6 = \boxed{36}$. Check the answer: The reversed number is $10(6) + 3 = 63$. The sum is $36 + 63 = 99$ and the difference is $63 - 36 = 27$.

PROBLEM #47

At a movie theater, the ratio of children's ticket sales to the total ticket sales is 7:9. The movie theater sold 360 more children's tickets than adult tickets. How many children's tickets and how many adult tickets did the movie theater sell?

SOLUTION #47

Fewer adult tickets were sold than children's tickets.

$$x = \text{the number of adult tickets sold}$$

The movie theater sold 360 more children's tickets than adult tickets.

$$x + 360 = \text{the number of children's tickets sold}$$

We will need to work with the total number of tickets sold (based on how the ratio is worded). Add the number of children's tickets sold to the number of adult tickets sold to find the total number of tickets sold.

$$x + (x + 360) = 2x + 360 = \text{the total number of tickets sold}$$

The ratio of children's tickets sold to the total number of tickets sold is 7:9. This ratio can be expressed as the fraction $\frac{7}{9}$. Multiply the total number of tickets sold by this fraction to get the number of children's tickets sold. (It would be incorrect to multiply x by $\frac{7}{9}$ because x represents the number of adult tickets sold, but the denominator of the ratio involves the total number of tickets, not just the adult tickets.) Recall that the number of children's tickets sold equals $x + 360$.

$$\frac{7}{9}(2x + 360) = x + 360$$

Multiply both sides of the equation by 9.

$$7(2x + 360) = 9(x + 360)$$

Distribute the 7 and the 9.

$$14x + 2520 = 9x + 3240$$

$$5x = 720$$

$$x = \frac{720}{5} = 144$$

The theater sold $x = \boxed{144}$ adult tickets and $x + 360 = 144 + 360 = \boxed{504}$ children's tickets. Check the answers: The total number of tickets sold is $144 + 504 = 648$, and the ratio of children's tickets sold to the total number of tickets sold equals $\frac{504}{648} = \frac{504 \div 72}{648 \div 72} = \frac{7}{9}$.

PROBLEM #48

Chang's purse just has dimes and Tammy's purse just has nickels. Chang has four times as much money as Tammy. If Chang gives 51 dimes to Tammy, then Tammy will have four times as much money as Chang. How many dimes did Chang have and how many nickels did Tammy have in the beginning?

SOLUTION #48

Originally, Tammy had fewer nickels than Chang had dimes.

$x =$ the number of nickels that Tammy had in the beginning

Multiply the number of nickels by 5 to determine the value of Tammy's money.

$5x =$ the amount of money that Tammy had in the beginning

Originally, Chang had four times as much money as Tammy. Multiply the value of Tammy's money by 4 to determine the value of Chang's money in the beginning.

$20x =$ the amount of money that Chang had in the beginning

Divide the value of Chang's money by 10 to determine how many dimes she had.

$\frac{20x}{10} = 2x =$ the number of dimes that Chang had in the beginning

The problem asks about Chang giving 51 dimes to Tammy. The value of 51 dimes is $(51)(10) = 510$ cents. Adjust Chang's and Tammy's amounts of money by 510 cents.

$5x + 510 =$ the amount of money that Tammy had after receiving 51 dimes

$20x - 510 =$ the amount of money that Chang had after giving 51 dimes

Afterward, Tammy has four times as much money as Chang.

$$5x + 510 = 4(20x - 510)$$
$$5x + 510 = 80x - 2040$$
$$510 + 2040 = 80x - 5x$$
$$2550 = 75x$$
$$\frac{2550}{75} = 34 = x$$

In the beginning, Tammy had $x = \boxed{34}$ nickels and Chang had $2x = 2(34) = \boxed{68}$ dimes. Check the answers: In the beginning, Tammy had $(34)(5) = 170$ cents and Chang had $(68)(10) = 680$ cents. In the beginning, $(170)(4) = 680$. After Chang gave 510 cents to Tammy, Tammy had $170 + 510 = 680$ cents and Chang had $680 - 510 = 170$ cents. Afterward, $680 = (170)(4)$.

PROBLEM #49

Rachel and Karen went to a store where there was no sales tax. Rachel paid \$5.30 for eight apples and three oranges. Karen paid \$1.80 for one apple and two oranges. How much did the store charge for apples and how much did the store charge for oranges?

SOLUTION #49

We will define two variables and solve this problem using two equations. (This is convenient, but not necessary. If you define y to be the price of an orange in cents, from the third sentence you could determine that the number of apples equals $180 - 2y$. Then you would not need to use x,)

$$x = \text{the price of one apple in cents}$$

$$y = \text{the price of one orange in cents}$$

Rachel paid \$5.30 for eight apples and three oranges. Multiply by 100 to get 530 cents.

$$8x + 3y = 530 \quad \text{(first equation)}$$

Karen paid \$1.80 for one apple and two oranges. Multiply by 100 to get 180 cents.

$$x + 2y = 180 \quad \text{(second equation)}$$

Subtract $2y$ from both sides of the second equation.

$$x = 180 - 2y$$

Substitute $180 - 2y$ in place of x in the first equation.

$$8(180 - 2y) + 3y = 530$$

$$1440 - 16y + 3y = 530$$

$$1440 - 13y = 530$$

$$1440 - 530 = 13y$$

$$910 = 13y$$

$$\frac{910}{13} = 70 = y$$

$$x = 180 - 2y = 180 - 2(70) = 180 - 140 = 40$$

The store charged $x = \boxed{40}$ cents for each apple and $y = \boxed{70}$ cents for each orange. Check the answers: The cost for 8 apples and 3 oranges is $8(\$0.40) + 3(\$0.70) = \$3.20 + \$2.10 = \$5.30$, and the cost for 1 apple and 2 oranges is $\$0.40 + 2(\$0.70) = \$0.40 + \$1.40 = \$1.80$.

PROBLEM #50

A man paid $352 for a rug, including 10% sales tax. The rug was on sale. The regular price of the rug was $400. What percentage discount did the man receive?

SOLUTION #50

$$x = \text{the discount expressed as a decimal}$$

Here is how the purchase works:

1. Begin with the regular price, which is \$400.
2. Multiply the regular price (\$400) by the decimal form of the discount (x) to get $400x$. This is the amount of the discount in dollars.
3. Subtract the discount from the regular price to get $400 - 400x$, which could also be expressed as $400(1 - x)$. This is the sale price.
4. Figure the tax. Divide 10% by 100% to convert the tax to a decimal: $\frac{10\%}{100\%} = 0.1$. Multiply 0.1 by the sale price to get $(0.1)(400)(1 - x) = 40(1 - x)$. This is the amount of the tax.
5. Add the tax to the sale price to get the purchase price. The purchase price is $400(1 - x) + 40(1 - x) = 400 - 400x + 40 - 40x = 440 - 440x$, which can also be expressed as $440(1 - x)$. According to the problem, the purchase price was \$352. Set the purchase price equal to 352 dollars.

$$440(1 - x) = 352$$

$$1 - x = \frac{352}{440} = \frac{352 \div 88}{440 \div 88} = \frac{4}{5} = 0.8$$

$$1 = 0.8 + x$$

$$1 - 0.8 = x$$

$$0.2 = x$$

The discount was $x = 0.2 = \boxed{20\%}$. Check the answer: 20% of \$400 is 0.2(\$400) = \$80, such that 20% off of \$400 equals \$400 − \$80 = \$320. The tax is 10% of \$320, which is 0.1(\$320) = \$32. Add the tax of \$32 to the sale price of \$320 to determine the purchase price: \$320 + \$32 = \$352.

PROBLEM #51

A man invested one-fourth of his money in stocks and put the rest of his money in a savings account. The stocks paid a return of 9% and the savings earned 2% interest. The total return was $600. How much was the total investment?

SOLUTION #51

$$x = \text{the amount invested in stocks}$$

The man invested one-fourth of his money in stocks and put the rest of his money in a savings account. This means that he invested three-fourths of his money in savings. Since x was invested in stocks, it follows that $3x$ was put into a savings account. The total investment is $x + 3x = 4x$. Observe that x is one-fourth of $4x$ and that $3x$ is three-fourths of $4x$.

$$3x = \text{the amount invested in a savings account}$$

The interest earned (I) equals the principal invested (P) times the interest rate (r): $I = Pr$. Apply this formula to each investment. Divide by 100% to convert 9% and 2% into decimals: $\frac{9\%}{100\%} = 0.09$ and $\frac{2\%}{100\%} = 0.02$.

$$0.09x = \text{the interest earned from the stocks at 9\%}$$

$$0.02(3x) = 0.06x = \text{the interest earned from the savings account at 2\%}$$

The total return was $600. Add the two returns together.

$$0.09x + 0.06x = 600$$

Multiply by 100 to remove all of the decimals.

$$9x + 6x = 60{,}000$$

$$15x = 60{,}000$$

$$x = \frac{60{,}000}{15} = 4000$$

The man invested $x = 4000$ dollars in stocks and $3x = 3(4000) = 12{,}000$ dollars in a savings account. The total investment was $4000 + 12{,}000 = \boxed{16{,}000}$ dollars (which equals $4x$). Check the answer: The total return is $0.09(\$4000) + 0.02(\$12{,}000) = \$360 + \$240 = \$600$.

PROBLEM #52

A woman travels in a motorboat along a river downstream from one town to another in 90 minutes. When she returns along the same route upstream, it takes 2 hours. The river current is 8 mph. What would be the speed of the motorboat in still water?

SOLUTION #52

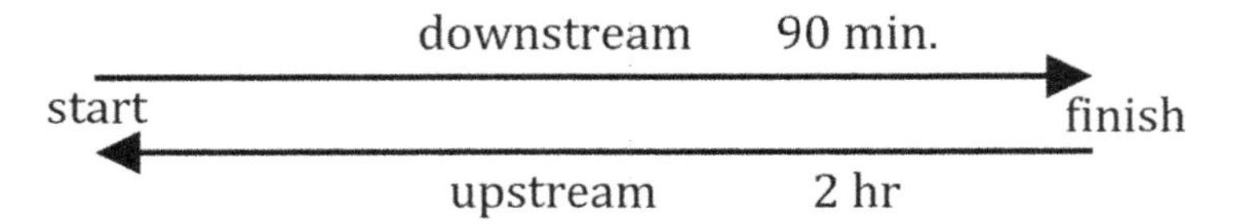

$r =$ the speed of the motorboat in still water in miles per hour (mph)

The motorboat travels faster downstream than upstream. Going downstream, add the river current (8 mph) to the boat speed. Going upstream, subtract the river current from the boat speed. Note that one time is given in minutes, whereas all other times include hours. You need to convert 90 minutes to hours in order to have consistent units: 90 min. = 1.5 hr.

	rate (mph)	time (hr)	distance (miles)
downstream	$r + 8$	1.5	$1.5(r + 8)$
upstream	$r - 8$	2	$2(r - 8)$

The boat travels the same distance each way. Set the distances equal to each other.

$$1.5(r + 8) = 2(r - 8)$$

$$1.5r + 12 = 2r - 16$$

Multiply both sides of the equation by 2 in order to remove the decimal point.

$$3r + 24 = 4r - 32$$

$$24 + 32 = 4r - 3r$$

$$56 = r$$

The motorboat would have a speed of $r = \boxed{56}$ mph in still water. Check the answer: The distance downstream is $1.5(r + 8) = 1.5(56 + 8) = 1.5(64) = 96$ miles and the distance upstream is $2(r - 8) = 2(56 - 8) = 2(48) = 96$ miles.

PROBLEM #53

How much pure sulfuric acid must be added to 5 quarts of a solution of 80% sulfuric acid in order to strengthen the solution to 90%?

SOLUTION #53

x = the volume of pure sulfuric acid in quarts that needs to be added

The amount of pure sulfuric acid (P) equals the decimal value of the percentage (c) times the volume of the mixture (M): $P = cM$. Apply this formula to each mixture. Divide the given percentages by 100% in order to convert them to decimals: $\frac{80\%}{100\%} = 0.8$ and $\frac{90\%}{100\%} = 0.9$. Organize the information in a table. The volume of the mixture is the sum of the volumes of the two solutions: $M_1 + M_2 = M_3$. Note that pure sulfuric acid is 100% sulfuric acid: As a decimal, 100% equals 1.

	80% H_2SO_4	pure (100% H_2SO_4)	90% H_2SO_4
c (decimal)	0.8	1	0.9
M (quarts)	5	x	$5 + x$
$P = cM$ (pure)	$(0.8)(5) = 4$	$1x = x$	$0.9(5 + x)$

The amount of pure stuff in the mixture is the sum of the amounts of the pure stuff in the two solutions: Add the values in the bottom row of the table.

$$4 + x = 0.9(5 + x)$$

$$4 + x = 4.5 + 0.9x$$

If you wish to remove the decimal, multiply both sides of the equation by 10.

$$40 + 10x = 45 + 9x$$

$$10x - 9x = 45 - 40$$

$$x = 5$$

The volume of pure sulfuric acid needed is $x = \boxed{5}$ quarts. Check the answer: $P_1 + P_2 = 4 + 5 = 9 = P_3$ agrees with $P_3 = c_3M_3 = 0.9(5 + x) = 0.9(5 + 5) = 0.9(10) = 9$.

PROBLEM #54

An empty bathtub could fill up in 12 minutes when the drain is plugged. The same empty bathtub takes an hour to fill up when the drain is unplugged. When the bathtub is full, the faucet is turned off, and the drain is unplugged, how long does it take for the bathtub to drain completely?

SOLUTION #54

$t =$ the time it takes for the bathtub to drain completely

Find the reciprocal of each given time in order to determine how much work is done each minute. Note that one hour equals 60 minutes.

	fill up (plugged)	drain	fill up (unplugged)
time to work (in minutes)	12	t	60
fraction completed per minute	$\frac{1}{12}$	$\frac{1}{t}$	$\frac{1}{60}$

Since the faucet and drain work against each other (instead of working together), we subtract the reciprocals from the bottom row of the table (instead of adding).

$$\frac{1}{12} - \frac{1}{t} = \frac{1}{60}$$

The lowest common denominator of 12, t, and 60 is $60t$. Multiply both sides by $60t$.

$$\frac{60t}{12} - \frac{60t}{t} = \frac{60t}{60}$$

$$5t - 60 = t$$

$$5t - t = 60$$

$$4t = 60$$

$$t = \frac{60}{4} = 15$$

The bathtub drains completely in $t = \boxed{15}$ minutes. Check the answer: To find the fraction of the work done, divide 60 minutes by the time it would take to work alone. The fraction of the work done by the faucet is $60 \div 12$ and the fraction of the work done by the drain is $60 \div 15$. When you subtract the fraction for the drain (since it works against the faucet), the result must be one (because all of the work gets done): $\frac{60}{12} - \frac{60}{15} = 5 - 4 = 1$. (Each "fraction" is bigger than one because if the faucet or drain worked alone, it would finish in less than 60 minutes.)

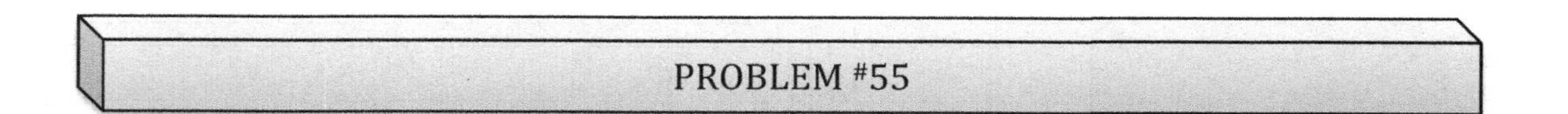

The system below is in static equilibrium. How far is the center of each box from the fulcrum?

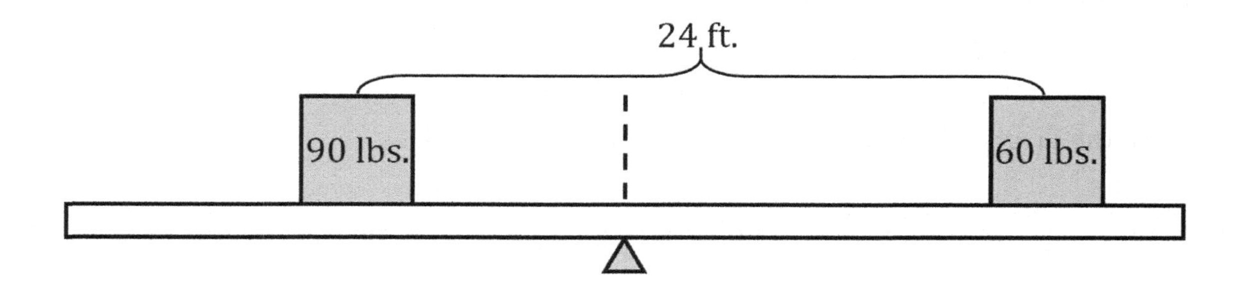

SOLUTION #55

r_1 = the distance from the fulcrum to the center of the left box

r_2 = the distance from the fulcrum to the center of the right box

The system is in static equilibrium. Apply the torque equation.

$$w_1 r_1 = w_2 r_2$$

$$90r_1 = 60r_2 \quad \text{(first equation)}$$

From the diagram, we see that r_1 and r_2 add up to 24 feet.

$$r_1 + r_2 = 24 \quad \text{(second equation)}$$

Isolate one of the unknowns. Subtract r_1 from both sides of the second equation.

$$r_2 = 24 - r_1$$

Substitute this expression for r_2 into the first equation.

$$90r_1 = 60(24 - r_1)$$

$$90r_1 = 1440 - 60r_1$$

$$90r_1 + 60r_1 = 1440$$

$$150r_1 = 1440$$

$$r_1 = \frac{1440}{150} = 9.6$$

The center of the box on the left side of the fulcrum is $r_1 = \boxed{9.6}$ feet from the fulcrum and the center of the box on the right side of the fulcrum is $r_2 = 24 - r_1 = 24 - 9.6 = \boxed{14.4}$ feet from the fulcrum. Check the answers: The two torques are $(90)(9.6) = 864$ and $(60)(14.4) = 864$.

Note: Don't try to guess the answers based on how far the distances look in the diagram because the diagram is not drawn to scale.

PROBLEM #56

Originally, the ratio of the width of a rectangle to the height of the rectangle is 4:3. The rectangle is altered such that its width and height each increase by 5 inches. The area of the new rectangle is greater than the area of the old rectangle by 200 square inches. What are the width and height of the original rectangle?

SOLUTION #56

$w =$ the width of the original rectangle in inches

$h =$ the height of the original rectangle in inches

Originally, the ratio of the width of the rectangle to the height of the rectangle is 4:3. This ratio can be expressed as the fraction $\frac{4}{3}$. Multiply the height by this fraction in order to get the width.

$$w = \frac{4}{3}h$$

The width and height each increase by 5 inches.

$w + 5 =$ the width of the altered rectangle in inches

$h + 5 =$ the height of the altered rectangle in inches

The area of a rectangle equals the width times the height. Apply this to the original rectangle and to the altered rectangle.

$wh =$ the area of the original rectangle

$(w + 5)(h + 5) = wh + 5w + 5h + 25 =$ the area of the altered rectangle

The new area is greater than the old area by 200 square inches.

$$wh + 5w + 5h + 25 = wh + 200 \quad \text{(subtract } wh \text{ and 25 from both sides)}$$

$$5w + 5h = 175 \quad \left(\text{recall that } w = \frac{4}{3}h\right)$$

$$5\left(\frac{4}{3}h\right) + 5h = 175 \quad \text{(multiply both sides by 3)}$$

$$20h + 15h = 525$$

$$35h = 525$$

$$h = \frac{525}{35} = 15$$

The height of the original rectangle is $h = \boxed{15}$ inches and the width is $w = \frac{4}{3}h = \frac{4}{3}(15) = \frac{60}{3} = \boxed{20}$ inches. Check the answers: The original area is $wh = (20)(15) = 300$ in.^{2} and the new area is $(w + 5)(h + 5) = (25)(20) = 500$ in.^{2} The new area is greater than the old area by $500 - 300 = 200$ square inches.

PROBLEM #57

Originally, Natalie had three times as many bracelets as Annette, and Annette had 12 more bracelets than Linda. After Natalie gave 24 bracelets to Linda and Natalie gave 12 bracelets to Annette, all three girls had the same number of bracelets. How many bracelets did each girl have in the beginning?

SOLUTION #57

In the beginning, Linda had fewer bracelets than Natalie or Annette.

$x =$ the number of bracelets that Linda had to begin with

In the beginning, Annette had 12 more bracelets than Linda.

$x + 12 =$ the number of bracelets that Annette had to begin with

In the beginning, Natalie had three times as many bracelets as Annette. It follows that:

$3(x + 12) = 3x + 36 =$ the number of bracelets that Natalie had to begin with

Natalie gave 24 bracelets to Linda and Natalie gave 12 bracelets to Annette. Subtract 36 bracelets from Natalie (since $24 + 12 = 36$), add 24 bracelets to Linda, and add 12 bracelets to Annette.

$(3x + 36) - 36 = 3x =$ the number of bracelets that Natalie has now

$x + 24 =$ the number of bracelets that Linda has now

$(x + 12) + 12 = x + 24 =$ the number of bracelets that Annette has now

After Natalie gave bracelets to the other girls, they have the same number of bracelets now. Set the current number of bracelets for Natalie and the others equal.

$$3x = x + 24$$

Combine like terms. Subtract x from both sides.

$$3x - x = 24$$

$$2x = 24$$

$$x = \frac{24}{2} = 12$$

In the beginning, Linda had $x = \boxed{12}$ bracelets, Annette had $x + 12 = 12 + 12 = \boxed{24}$ bracelets, and Natalie had $3(x + 12) = 3(12 + 12) = 3(24) = \boxed{72}$ bracelets. Check the answers: After Natalie gave 36 bracelets to the other girls, Natalie had $72 - 36 = 36$ bracelets. After Linda received 24 bracelets, Linda had $12 + 24 = 36$ bracelets. After Annette received 12 bracelets, Annette had $24 + 12 = 36$ bracelets.

PROBLEM #58

Nine years ago, Miguel was twice as old as Carlos and Sylvia was four times as old as Miguel. In two years, the sum of their ages will be 99. What are their ages now?

SOLUTION #58

Carlos is younger than Miguel and Sylvia. Note that we are defining our variable to be 9 years ago (instead of now).

$$x = \text{Carlos's age 9 years ago}$$

Nine years ago, Miguel was twice as old as Carlos and Sylvia was four times as old as Miguel.

$$2x = \text{Miguel's age 9 years ago}$$
$$4(2x) = 8x = \text{Sylvia's age 9 years ago}$$

The second sentence refers to their ages two years from now. Since we defined our ages to be 9 years ago, we need to add 11 years to those ages to make them 2 years from now. That is, it takes 9 years to get to now and an additional 2 years makes 11.

$$x + 11 = \text{Carlos's age 2 years from now}$$
$$2x + 11 = \text{Miguel's age 2 years from now}$$
$$8x + 11 = \text{Sylvia's age 2 years from now}$$

In two years, the sum of their ages will be 99.

$$(x + 11) + (2x + 11) + (8x + 11) = 99$$
$$11x + 33 = 99$$
$$11x = 66$$
$$x = \frac{66}{11} = 6$$

The question asks for their current ages. Since we defined x to be 9 years ago, we must add 9 to their ages to find their current ages. Carlos is $x + 9 = 6 + 9 = \boxed{15}$ years old now, Miguel is $2x + 9 = 2(6) + 9 = \boxed{21}$ years old now, and Sylvia is $8x + 9 = 8(6) + 9 = \boxed{57}$ years old now. Check the answers: Nine years ago, Carlos was $15 - 9 = 6$, Miguel was $21 - 9 = 12$, and Sylvia was $57 - 9 = 48$. Nine years ago, Miguel (12) was twice as old as Carlos (6) and Sylvia (48) was four times as old as Miguel (12). In two years, Carlos will be $15 + 2 = 17$, Miguel will be $21 + 2 = 23$, and Sylvia will be $57 + 2 = 59$. In two years, the sum of their ages will be $17 + 23 + 59 = 99$.

PROBLEM #59

The sum of the digits of a three-digit number is 18. The tens digit is 4 more than the hundreds digit. When the digits of the three-digit number are reversed, the reversed number is smaller than the original number by 99. What is the number?

SOLUTION #59

$$x = \text{the units digit of the original number}$$

$$y = \text{the hundreds digit of the original number}$$

The tens digit is 4 more than the hundreds digit.

$$y + 4 = \text{the tens digit of the original number}$$

The sum of the digits of the three-digit number is 18.

$$y + (y + 4) + x = 18$$

$$2y + x = 14 \quad \text{(first equation)}$$

To make the original number, multiply the hundreds digit by 100, the tens digit by 10, and the units digit by 1.

$$100y + 10(y + 4) + 1x = 110y + x + 40 = \text{the original number}$$

To make the reversed number, swap the hundreds digit with the units digit.

$$y = \text{the units digit of the reversed number}$$

$$x = \text{the hundreds digit of the reversed number}$$

$$100x + 10(y + 4) + 1y = 100x + 11y + 40 = \text{the reversed number}$$

The reversed number is smaller than the original number by 99.

$$100x + 11y + 40 = (110y + x + 40) - 99$$

$$99x = 99y - 99 \quad \text{(divide by 99)}$$

$$x = y - 1 \quad \text{(second equation)}$$

Substitute the second equation into the first equation.

$$2y + (y - 1) = 14$$

$$3y = 15$$

$$y = \frac{15}{3} = 5$$

The hundreds digit is $y = 5$, the tens digit is $y + 4 = 5 + 4 = 9$, and the units digit is $x = y - 1 = 5 - 1 = 4$. The original number is $100(5) + 10(9) + 1(4) = 500 + 90 + 4 = \boxed{594}$. Check the answer: The digits add up to $5 + 9 + 4 = 18$ and the tens digit is 4 more than the hundreds digit (since $9 - 5 = 4$). The reversed number is 495. The original number minus the reversed number is $594 - 495 = 99$.

PROBLEM #60

A gumball machine has red, green, and blue gumballs in the ratio of 7:3:8. There are 450 gumballs in the machine. How many gumballs of each color are there?

SOLUTION #60

There are fewer green gumballs than red or blue gumballs.

$$x = \text{the number of green gumballs}$$

The red, green, and blue gumballs come in the ratio of 7:3:8. This means that the ratio of red gumballs to green gumballs is 7:3, and the ratio of blue gumballs to green gumballs is 8:3 (it is equivalent to saying that the ratio of green gumballs to blue gumballs is 3:8, but we wish to put green in the denominator). These ratios can be expressed as the fractions $\frac{7}{3}$ and $\frac{8}{3}$. Multiply the number of green gumballs by these fractions to get the number of red gumballs and the number of blue gumballs.

$$\frac{7x}{3} = \text{the number of red gumballs}$$

$$\frac{8x}{3} = \text{the number of blue gumballs}$$

There are 450 gumballs in the machine. Add up the numbers of gumballs.

$$x + \frac{7x}{3} + \frac{8x}{3} = 450$$

Multiply both sides of the equation by 3.

$$3x + 7x + 8x = 1350$$

$$18x = 1350$$

$$x = \frac{1350}{18} = 75$$

There are $x = \boxed{75}$ green gumballs, $\frac{7x}{3} = \frac{7(75)}{3} = \frac{525}{3} = \boxed{175}$ red gumballs, and $\frac{8x}{3} = \frac{8(75)}{3} = \frac{600}{3} = \boxed{200}$ blue gumballs. Check the answers: The total number of gumballs is $75 + 175 + 200 = 450$. The ratio of red gumballs to green gumballs equals $\frac{175}{75} = \frac{175 \div 25}{75 \div 25} = \frac{7}{3}$, and the ratio of green gumballs to blue gumballs equals $\frac{75}{200} = \frac{75 \div 25}{200 \div 25} = \frac{3}{8}$ (so the ratio of blue gumballs to green gumballs is $\frac{8}{3}$). The ratio of red to green to blue gumballs is 7:3:8 (which combines the ratios 7:3 and 3:8 together, with 3 being the middleman).

PROBLEM #61

Joey and Oscar only have five-dollar ($5) and ten-dollar ($10) bills in their wallets. Joey has four times as many ten-dollar ($10) bills as Oscar, but Oscar has eight more five-dollar ($5) bills than Joey. If they combine their money together, Joey and Oscar have $560. If they keep their money separate, Joey has $200 more than Oscar. How much money does each man have in his wallet?

SOLUTION #61

$$x = \text{the number of five-dollar (\$5) bills in Oscar's wallet}$$
$$y = \text{the number of ten-dollar (\$10) bills in Oscar's wallet}$$

Joey has four times as many $10 bills as Oscar. Since Oscar has y $10 bills:

$$4y = \text{the number of ten-dollar (\$10) bills in Joey's wallet}$$

Oscar has eight more $5 bills than Joey. This means that Joey has eight fewer $5 bills than Oscar. Since Oscar has x $5 bills:

$$x - 8 = \text{the number of five-dollar (\$5) bills in Joey's wallet}$$

If they combine their money together, Joey and Oscar have $560. Add up the money that each man has. Multiply the number of $5 bills by 5 and the number of $10 bills by 10 in order to determine the value of the money in dollars.

$$5x + 5(x - 8) + 10y + 10(4y) = 560$$
$$5x + 5x - 40 + 10y + 40y = 560$$
$$10x + 50y = 600 \quad \text{(divide by 10)}$$
$$x + 5y = 60$$
$$x = 60 - 5y \quad \text{(first equation)}$$

Joey has $200 more than Oscar. Determine how much money Oscar has and how much money Joey has, and add $200 to Oscar's money to get Joey's money.

$$5x + 10y + 200 = 5(x - 8) + 10(4y)$$
$$5x + 10y + 200 = 5x - 40 + 40y \quad \text{(note that } 5x \text{ cancels out)}$$
$$240 = 30y$$
$$\frac{240}{30} = 8 = y \quad \text{(second equation)}$$

Plug $y = 8$ into the first equation.

$$x = 60 - 5y = 60 - 5(8) = 60 - 40 = 20$$

Oscar has $x = 20$ five-dollar bills and $y = 8$ ten-dollar bills. Oscar has $5x + 10y = 5(20) + 10(8) = 100 + 80 = \boxed{180}$ dollars. Joey has $x - 8 = 20 - 8 = 12$ five-dollar bills and $4y = 4(8) = 32$ ten-dollar bills. Joey has $5(12) + 10(32) = 60 + 320 = \boxed{380}$ dollars. Check the answers: $\$380 + \$180 = \$560$ and $\$380 - \$180 = \$200$.

PROBLEM #62

Gary, Keith, and Olivia went to a restaurant where there was no sales tax. Gary paid $4.25 for large fries and a large soda. Keith paid $8.00 for a hamburger, large fries, and a large soda. Olivia paid $5.50 for a hamburger and a large soda. How much did the restaurant charge for hamburgers, for large fries, and for large sodas?

SOLUTION #62

$x =$ the price of one hamburger in cents

$y =$ the price of one order of large fries in cents

$z =$ the price of one large soda in cents

Multiply by 100 to make cents. Gary paid 425 cents for large fries and a large soda.

$$y + z = 425 \quad \text{(first equation)}$$

Keith paid $8.00 for a hamburger, large fries, and a large soda.

$$x + y + z = 800 \quad \text{(second equation)}$$

Olivia paid $5.50 for a hamburger and a large soda.

$$x + z = 550 \quad \text{(third equation)}$$

Isolate y in the first equation.

$$y = 425 - z$$

Replace y with $425 - z$ in the second equation. Note that z cancels in this equation.

$$x + 425 - z + z = 800$$

$$x + 425 = 800$$

$$x = 800 - 425 = 375$$

Plug $x = 375$ into the third equation.

$$375 + z = 550$$

$$z = 550 - 375 = 175$$

Plug $z = 175$ into the equation $y = 425 - z$ from earlier.

$$y = 425 - z = 425 - 175 = 250$$

The restaurant charged $x = \boxed{\$3.75}$ for one hamburger, $y = \boxed{\$2.50}$ for one order of large fries, and $z = \boxed{\$1.75}$ for one large soda. Check the answers: The cost for one large fries and one large soda is $\$2.50 + \$1.75 = \$4.25$. The cost for one hamburger, one large fries, and one large soda is $\$3.75 + \$2.50 + \$1.75 = \$8.00$. The cost for one hamburger and one large soda is $\$3.75 + \$1.75 = \$5.50$.

PROBLEM #63

A woman purchases groceries and household items from a store. Before tax, the cost of the groceries is \$25 more than the cost of the household items. She is charged 3% tax on the groceries and 7% tax on the household items. She paid a total of \$103.45. What was the cost of the groceries and the cost of the household items before tax?

SOLUTION #63

$x =$ the cost of the groceries in dollars before tax

$y =$ the cost of the household items in dollars before tax

Before applying tax, the cost of the groceries is \$25 more than the cost of the household items.

$$x = y + 25 \quad \text{(first equation)}$$

She is charged 3% tax on the groceries and 7% tax on the household items. Add 100% to get 103% and 107% (since the tax is in addition to the cost). Divide by 100% to convert these percents to decimals: $\frac{103\%}{100\%} = 1.03$ and $\frac{107\%}{100\%} = 1.07$. Multiply by these values to figure the cost of the groceries and household items after tax.

$1.03x =$ the cost of the groceries in dollars after tax

$1.07y =$ the cost of the household items in dollars after tax

The woman paid a total of \$103.45.

$$1.03x + 1.07y = 103.45$$

Multiply both sides of the equation by 100 in order to remove the decimals.

$$103x + 107y = 10{,}345 \quad \text{(second equation)}$$

Substitute the first equation (from earlier) into the second equation.

$$103(y + 25) + 107y = 10{,}345$$

$$103y + 2575 + 107y = 10{,}345$$

$$210y = 7770$$

$$y = \frac{7770}{210} = 37$$

Plug $y = 37$ into the first equation.

$$x = y + 25 = 37 + 25 = 62$$

Before tax, the cost of the groceries was $x = \boxed{62}$ dollars and the cost of the household items was $y = \boxed{37}$ dollars. Check the answers: 3% tax on \$62 is $(0.03)(\$62) = \$1.86$ and 7% tax on \$37 is $(0.07)(\$37) = \$2.59$. The total amount paid comes to $\$62 + \$37 + \$1.86 + \$2.59 = \$103.45$.

PROBLEM #64

An investor paid a total of $6000 for two different stocks. One stock made a profit of 8%, while the other stock suffered a loss of 12%. The net profit from the stocks was exactly zero. How much money was invested in each stock?

SOLUTION #64

$x =$ the amount invested in the stock that made a profit of 8%

The total amount invested was \$6000. Part of this was invested in the stock that made a profit of 8% and the remainder was invested in the stock that suffered a loss of 12%. The two investments add up to \$6000. It follows that:

$6000 - x =$ the amount invested in the stock that suffered a loss of 12%

The interest earned (I) equals the principal invested (P) times the interest rate (r): $I = Pr$. Apply this formula to each investment. Divide by 100% to convert 8% and 12% into decimals: $\frac{8\%}{100\%} = 0.08$ and $\frac{12\%}{100\%} = 0.12$.

$0.08x =$ the profit from the stock that gained 8%

$0.12(6000 - x) = 720 - 0.12x =$ the loss from the stock that lost 12%

The total return was \$0. The profit minus the loss equals \$0. (If they were both profits, we would add them, but one was a loss, so we subtract the loss. You might think of the loss as a negative profit.)

$$0.08x - (720 - 0.12x) = 0$$

Add $(720 - 0.12x)$ to both sides of the equation.

$$0.08x = (720 - 0.12x)$$

$$0.08x = 720 - 0.12x$$

Multiply both sides of the equation by 100 in order to remove all of the decimals.

$$8x = 72{,}000 - 12x$$

$$20x = 72{,}000$$

$$x = \frac{72{,}000}{20} = 3600$$

The investor purchased the stock that returned an 8% profit for $x = \boxed{3600}$ dollars and purchased the stock that suffered a 12% loss for $6000 - 3600 = \boxed{2400}$ dollars. Check the answers: The profit from the first stock is (0.08)(\$3600) = \$288. The loss from the second stock is (0.12)(\$2400) = \$288. The profit of \$288 minus the loss of \$288 yields a total return of \$288 − \$288 = \$0.

PROBLEM #65

Jasmine writes down three numbers on a piece of paper. The average of the smallest two numbers is 88, the average of all three numbers is 200, and the largest number minus the smallest number equals 383. What are the numbers?

SOLUTION #65

$$x = \text{the smallest number}$$
$$y = \text{the middle number}$$
$$z = \text{the largest number}$$

The largest number minus the smallest number equals 383.

$$z - x = 383 \quad \text{(first equation)}$$

The average of the smallest two numbers is 88. To find the average value of two numbers, add the numbers together and divide by two.

$$\frac{x + y}{2} = 88 \quad \text{(multiply by 2)}$$
$$x + y = 176 \quad \text{(second equation)}$$

The average of all three numbers is 200. To find the average value of three numbers, add the numbers together and divide by three.

$$\frac{x + y + z}{3} = 200 \quad \text{(multiply by 3)}$$
$$x + y + z = 600 \quad \text{(third equation)}$$

Plug the second equation into the third equation: Replace $x + y$ with 176.

$$176 + z = 600$$
$$z = 600 - 176 = 424$$

Plug $z = 424$ into the first equation.

$$424 - x = 383 \quad \text{(add } x \text{ and subtract 383)}$$
$$424 - 383 = 41 = x$$

Plug $x = 41$ into the second equation.

$$41 + y = 176$$
$$y = 176 - 41 = 135$$

The smallest number is $x = \boxed{41}$, the middle number is $y = \boxed{135}$, and the largest number is $z = \boxed{424}$. Check the answers: The average of the smallest two numbers is $\frac{41+135}{2} = \frac{176}{2} = 88$, the average of all three numbers is $\frac{41+135+424}{3} = \frac{600}{3} = 200$, and the largest number minus the smallest number equals $424 - 41 = 383$.

PROBLEM #66

A boy and a dog are initially 2400 feet apart. The dog begins moving 25 ft./s towards the boy. Eight seconds later, the boy begins moving 15 ft./s towards the dog. Where do the boy and dog meet?

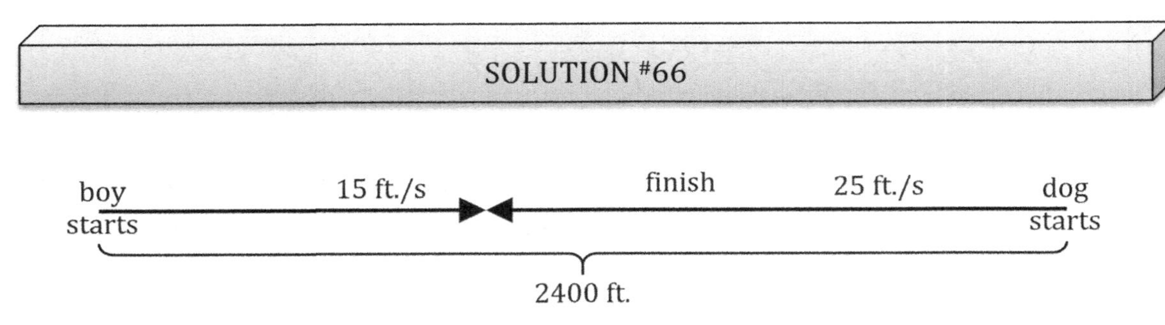

t = the amount of time in seconds that has passed since the dog began moving

The boy begins moving eight seconds after the dog began moving. This means that the boy spends less time moving.

$t - 8$ = the amount of time that has passed since the boy began moving

Distance (d) equals rate (r) times time (t): $d = rt$. The rates are 15 ft./s and 25 ft./s. Apply the rate equation to the boy and the dog. Organize the information in a table.

	rate (ft./s)	time (s)	distance (ft.)
boy	15	$t - 8$	$15(t - 8)$
dog	25	t	$25t$

The boy and the dog are initially 2400 feet apart. The distances add up to 2400 ft.

$$15(t - 8) + 25t = 2400$$
$$15t - 120 + 25t = 2400$$
$$40t = 2520$$
$$t = \frac{2520}{40} = 63$$

This isn't the final answer because the question asked where they meet (not when they meet). Use the rate equation to determine how far the boy and dog have traveled.

$$d_b = r_b t_b = 15(t - 8) = 15(63 - 8) = 15(55) = 825$$
$$d_d = r_d t_d = 25t = 25(63) = 1575$$

The boy has traveled $d_b = \boxed{825}$ feet and the dog has traveled $d_d = \boxed{1575}$ feet from their respective starting points when they meet. Check the answer: The total distance traveled is $825 + 1575 = 2400$ feet.

PROBLEM #67

A 600 mL solution of hydrochloric acid (of unknown concentration) is mixed with a 300 mL solution of 70% hydrochloric acid. The resulting mixture is 40% hydrochloric acid. What percent of the 600 mL solution is hydrochloric acid?

SOLUTION #67

x = the percent of hydrochloric acid in the 600 mL solution, expressed as a decimal

The amount of hydrochloric acid (P) equals the decimal value of the percentage (c) times the volume of the solution (M): $P = cM$. Apply this formula to each solution. Divide the given percentages by 100% in order to convert them to decimals: $\frac{70\%}{100\%} = 0.7$ and $\frac{40\%}{100\%} = 0.4$. Organize the information in a table. The volume of the mixture is the sum of the volumes of the two solutions: $M_1 + M_2 = M_3$.

	x HCl	70% HCl	40% HCl
c (decimal)	x	0.7	0.4
M (milliliters)	600	300	$600 + 300 = 900$
$P = cM$ (pure)	$600x$	$(0.7)(300) = 210$	$(0.4)(900) = 360$

The amount of pure stuff in the mixture is the sum of the amounts of the pure stuff in the two solutions: Add the values in the bottom row of the table.

$$600x + 210 = 360$$

$$600x = 150$$

$$x = \frac{150}{600} = \frac{150 \div 150}{600 \div 150} = \frac{1}{4} = 0.25$$

Multiply by 100% to convert this to a percent: $x = (0.25)100\% = \boxed{25\%}$. Check the answer: $P_3 = c_3 M_3 = (0.4)(900) = 360$ agrees with $P_1 + P_2 = (600)(0.25) + 210 = 150 + 210 = 360 = P_3$.

PROBLEM #68

Two monkeys are using buckets and a nearby lake to fill up a tub of water. If the tub didn't have a leak, one of the monkeys could fill up the tub in 6 hours if he worked alone, and the other monkey could fill up the tub in 8 hours if he worked alone. If the tub were full of water and the monkeys weren't working, the leak would cause the tub to completely drain in 12 hours. If the two monkeys work together, how long will it take for them to fill up the tub (assuming that the monkeys don't repair the leak)?

SOLUTION #68

t = the time it takes for the monkeys to fill the tub (while there is a leak)
Find the reciprocal of each given time in order to determine how much work is done each hour.

	1st monkey	2nd monkey	leak	all together
time to work (in hours)	6	8	12	t
fraction completed per hour	$\frac{1}{6}$	$\frac{1}{8}$	$\frac{1}{12}$	$\frac{1}{t}$

The monkeys work together, but the leak works against the monkeys. Therefore, we add the reciprocals for the monkeys, but subtract the reciprocal for the leak.

$$\frac{1}{6} + \frac{1}{8} - \frac{1}{12} = \frac{1}{t}$$

The lowest common denominator of 6, 8, 12, and t is $24t$. Multiply both sides by $24t$.

$$\frac{24t}{6} + \frac{24t}{8} - \frac{24t}{12} = \frac{24t}{t}$$

$$4t + 3t - 2t = 24$$

$$5t = 24$$

$$t = \frac{24}{5} = 4.8$$

The monkeys fill the tub completely in $t = \boxed{4.8}$ hours despite the fact that the tub is leaking. Check the answer: To find the fraction of the work done, divide 4.8 hours by the time it would take to work alone. For the first monkey, the fraction of the work done is $4.8 \div 6$, and for the second monkey the fraction of the work done is $4.8 \div 8$. For the leak, the fraction of the work is $4.8 \div 12$. If you add the fractions for the monkeys together and subtract the fraction for the leak (because it works against the monkeys), the sum must add up to one (because all of the work gets done): $\frac{4.8}{6} + \frac{4.8}{8} - \frac{4.8}{12} = 0.8 + 0.6 - 0.4 = 1$.

PROBLEM #69

A girl cuts a square and a circle out of a sheet of paper such that the square and the circle have the exact same area. What is the ratio of the diameter of the circle to the length of the square?

SOLUTION #69

$$L = \text{the length of the square}$$
$$r = \text{the radius of the circle}$$
$$D = \text{the diameter of the circle}$$

The formula for the area of a square is $A = L^2$ (since all four sides of a square have the same length). The formula for the area of a circle is $A = \pi r^2$. The square and the circle have the same area. Set these areas equal to one another.

$$L^2 = \pi r^2$$

Squareroot both sides of the equation. Recall from algebra that $\sqrt{ab} = \sqrt{a}\sqrt{b}$.

$$\sqrt{L^2} = \sqrt{\pi}\sqrt{r^2}$$

Recall that $\sqrt{x^2} = x$. Also note that the order of multiplication doesn't matter.

$$L = r\sqrt{\pi}$$

The problem asks about diameter, not radius. Recall that radius is half the diameter: $r = \frac{D}{2}$. Replace the radius with $\frac{D}{2}$ in the previous equation.

$$L = \frac{D}{2}\sqrt{\pi} \quad \text{(multiply both sides by 2)}$$
$$2L = D\sqrt{\pi} \quad \left(\text{divide both sides by } \sqrt{\pi}\right)$$
$$\frac{2L}{\sqrt{\pi}} = D \quad \text{(divide both sides by } L\text{)}$$
$$\frac{2}{\sqrt{\pi}} = \frac{D}{L}$$

The ratio of the diameter to the length is $\frac{D}{L} = \boxed{\frac{2}{\sqrt{\pi}}} \approx \frac{2}{\sqrt{3.14159}} \approx \boxed{1.13}$. Note that many algebra teachers would want you to multiply by $\frac{\sqrt{\pi}}{\sqrt{\pi}}$ in order to rationalize the denominator, in which case the answer is $\frac{D}{L} = \frac{2}{\sqrt{\pi}}\frac{\sqrt{\pi}}{\sqrt{\pi}} = \boxed{\frac{2\sqrt{\pi}}{\pi}} \approx \frac{2\sqrt{3.14159}}{3.14159} \approx \boxed{1.13}$. Check the answer: $D = \frac{2L}{\sqrt{\pi}}$ and $r = \frac{D}{2} = \frac{L}{\sqrt{\pi}}$ such that $A = \pi r^2 = \pi\left(\frac{L}{\sqrt{\pi}}\right)^2 = \pi\frac{L^2}{\pi} = L^2$.

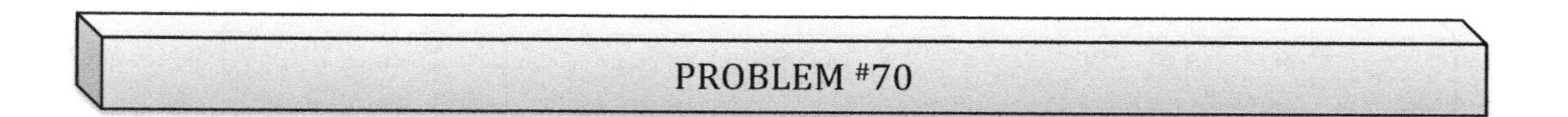

The mallet illustrated below consists of a 60-cm long handle and a 4-cm wide end. The mallet is in static equilibrium. The mallet has a total weight of 80 N (including both the handle and the end). If we cut the mallet into two pieces at the point where it is currently balanced (54 cm from the left end), how much will each end weigh?

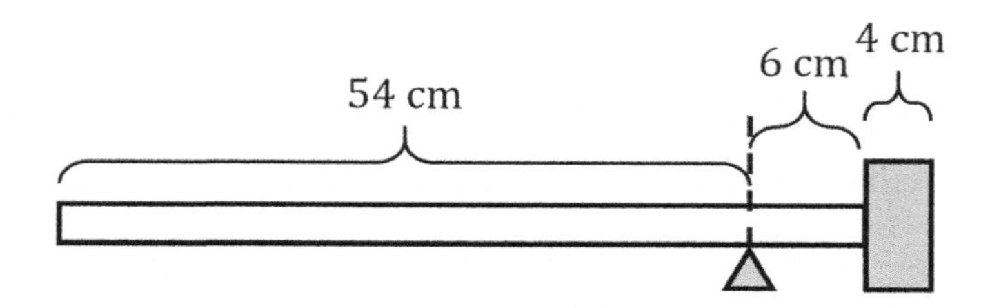

SOLUTION #70

w_1 = the weight of the left part of the handle (measuring 54 cm)

w_2 = the weight of the right part of the handle (measuring 6 cm)

w_3 = the weight of the end of the mallet (measuring 4 cm)

When a lever is in static equilibrium, the counterclockwise torques and clockwise torques are equal. Note that the weights are **not** equal. You can see this, for example, in Example 25 in Chapter 3. The right end (which includes the end of the mallet plus part of the handle) will weigh more than the left end because the right end is closer to the fulcrum. We will divide the mallet into three pieces as illustrated below.

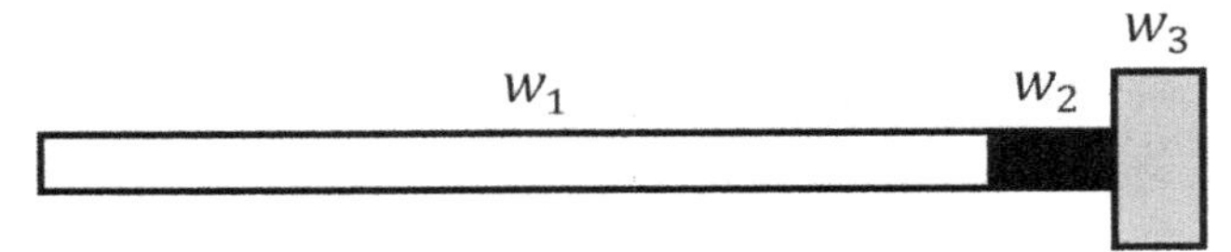

The handle is 60 cm long. The part of the handle called w_1 measures 54 cm long and the part of the handle called w_2 measures 6 cm long. Since w_1 is $\frac{54}{6} = 9$ times longer than w_2, w_1 must be 9 times heavier than w_2 (assuming that the handle is uniform).

$$w_1 = 9w_2 \quad \text{(first equation)}$$

The mallet has a total weight of 80 N.

$$w_1 + w_2 + w_3 = 80 \quad \text{(second equation)}$$

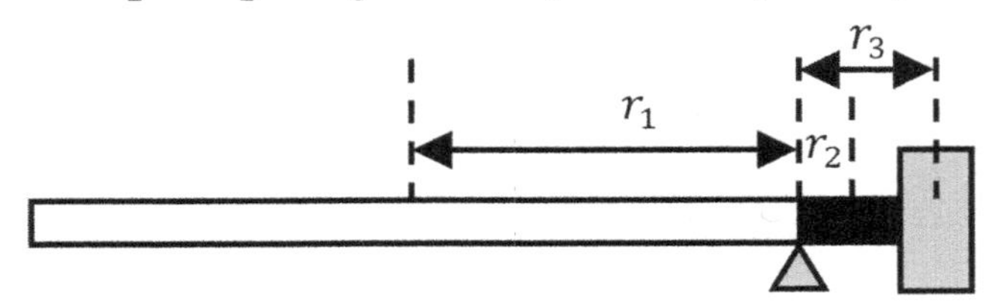

Set the sum of the counterclockwise torques equal to the sum of the clockwise torques, where each torque equals weight times lever arm. Compare the diagram above with the given diagram to see that $r_1 = \frac{54}{2} = 27$ cm, $r_2 = \frac{6}{2} = 3$ cm, and $r_3 = 6 + \frac{4}{2} = 8$ cm. Note that each r is measured from the center of each piece to the fulcrum.

$$w_1 r_1 = w_2 r_2 + w_3 r_3$$

$$27w_1 = 3w_2 + 8w_3 \quad \text{(third equation)}$$

Substitute the first equation into the second and third equations.

$$9w_2 + w_2 + w_3 = 80 \quad \to \quad 10w_2 + w_3 = 80$$

$$27(9w_2) = 3w_2 + 8w_3 \quad \to \quad 243w_2 = 3w_2 + 8w_3 \quad \to \quad 240w_2 = 8w_3 \quad \to \quad 30w_2 = w_3$$

Substitute $30w_2 = w_3$ into the equation $10w_2 + w_3 = 80$ above.

$$10w_2 + 30w_2 = 80$$

$$40w_2 = 80$$

$$w_2 = \frac{80}{40} = 2$$

Plug $w_2 = 2$ into the equation $w_1 = 9w_2$ (which was our first equation).

$$w_1 = 9w_2 = 9(2) = 18$$

Plug $w_1 = 18$ and $w_2 = 2$ into $w_1 + w_2 + w_3 = 80$ (which was our second equation).

$$18 + 2 + w_3 = 80$$

$$w_3 = 80 - 20 = 60$$

The left end (the part of the handle to the left of the fulcrum) has a weight equal to $w_1 = \boxed{18}$ N, and the right end (including both the end of the mallet and the part of the handle to the right of the fulcrum) has a combined weight equal to $w_2 + w_3 = 2 + 60 = \boxed{62}$ N. Check the answers: The counterclockwise torque is $(27)(18) = 486$ and the clockwise torque is $(3)(2) + (8)(60) = 6 + 480 = 486$. It should also make sense that the right end (weighing 62 N) is heavier than the left end (weighing 18 N), since the right end is closer to the fulcrum than the left end. (If you have any experience with a playground seesaw, you should know that the heavier person should sit closer to the fulcrum – or hinge – in order to balance the seesaw.)

THANK YOU

Thank you for reading:

INDEX

WAS THIS BOOK HELPFUL?

A great deal of effort and thought was put into this book, such as:

- Detailed explanations to help understand the ideas.
- Breaking down the solutions to help make the math easier to understand.
- Careful selection of problems for their instructional value.
- Ample examples to help illustrate the strategies and concepts.

If you appreciate the effort that went into making this book possible, there is a simple way that you could show it:

Please take a moment to post an honest review.

For example, you can review this book at Amazon.com or Barnes & Noble's website at BN.com.

Even a short review can be helpful and will be much appreciated. If you're not sure what to write, following are a few ideas, though it's best to describe what's important to you.

- Did you enjoy the selection of problems?
- Were you able to understand the solutions and explanations?
- Do you appreciate the handy formulas on the back cover of the print edition?
- How much did you learn from reading and using this workbook?
- Would you recommend this book to others? If so, why?

Do you believe that you found a mistake? Please email the author, Chris McMullen, at greekphysics@yahoo.com to ask about it. One of two things will happen:

- You might discover that it wasn't a mistake after all and learn why.
- You might be right, in which case the author will be grateful and future readers will benefit from the correction. Everyone is human.

ABOUT THE AUTHOR

Dr. Chris McMullen has over 20 years of experience teaching university physics in California, Oklahoma, Pennsylvania, and Louisiana. Dr. McMullen is also an author of math and science workbooks. Whether in the classroom or as a writer, Dr. McMullen loves sharing knowledge and the art of motivating and engaging students.

The author earned his Ph.D. in phenomenological high-energy physics (particle physics) from Oklahoma State University in 2002. Originally from California, Chris McMullen earned his Master's degree from California State University, Northridge, where his thesis was in the field of electron spin resonance.

As a physics teacher, Dr. McMullen observed that many students lack fluency in fundamental math skills. In an effort to help students of all ages and levels master basic math skills, he published a series of math workbooks on arithmetic, fractions, long division, algebra, trigonometry, and calculus entitled *Improve Your Math Fluency*. Dr. McMullen has also published a variety of science books, including introductions to basic astronomy and chemistry concepts in addition to physics workbooks.

Author, Chris McMullen, Ph.D.

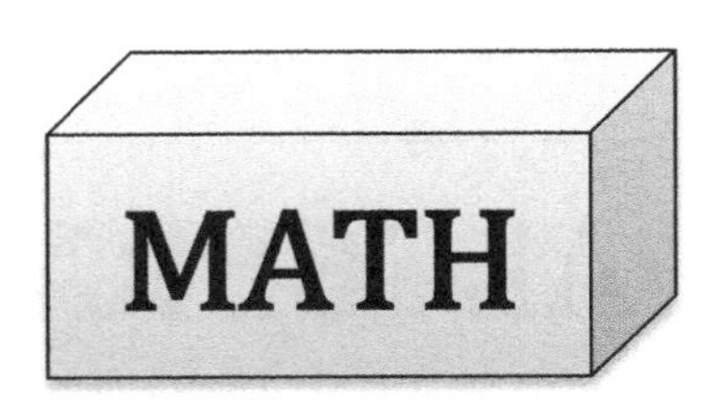

This series of math workbooks is geared toward practicing essential math skills:

- Algebra and trigonometry
- Calculus
- Fractions, decimals, and percentages
- Long division
- Multiplication and division
- Addition and subtraction

www.improveyourmathfluency.com

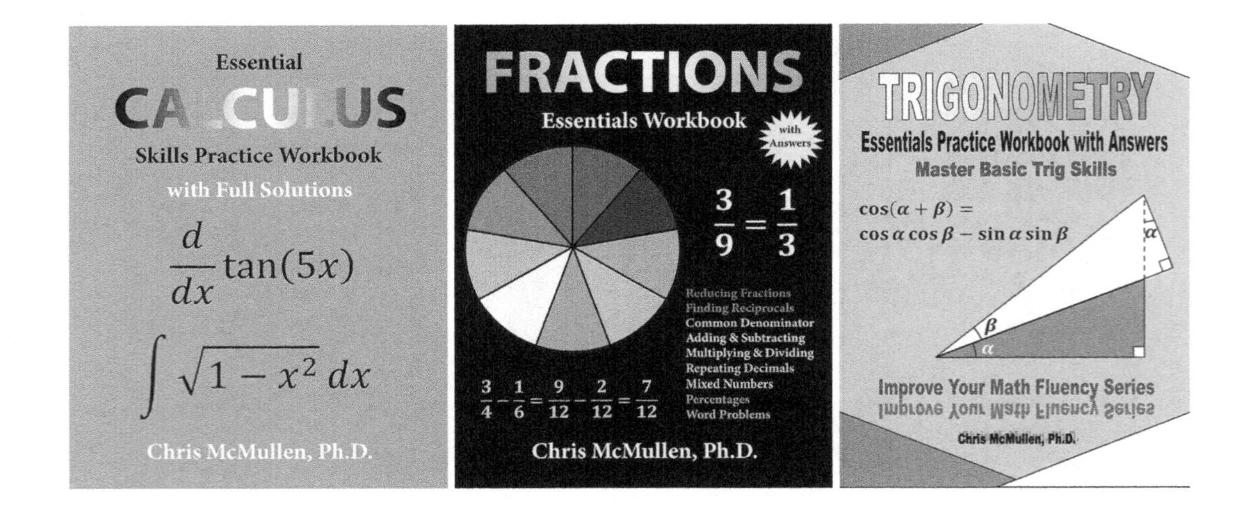

The author of this book, Chris McMullen, enjoys solving puzzles. His favorite puzzle is Kakuro (kind of like a cross between crossword puzzles and Sudoku). He once taught a three-week summer course on puzzles. If you enjoy mathematical pattern puzzles, you might appreciate:

300+ Mathematical Pattern Puzzles

Number Pattern Recognition & Reasoning

- Pattern recognition
- Visual discrimination
- Analytical skills
- Logic and reasoning
- Analogies
- Mathematics

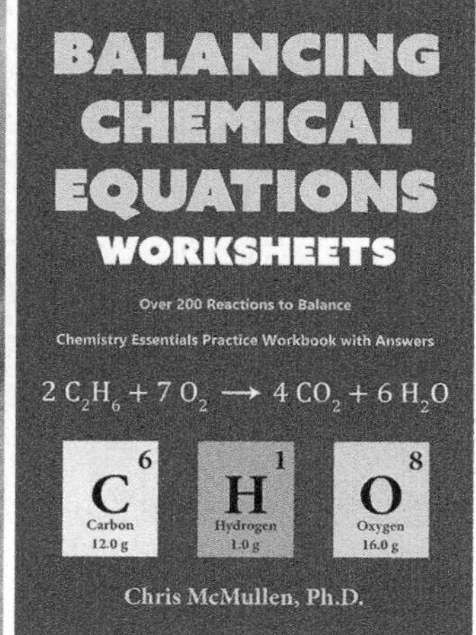

Dr. McMullen has published a variety of **science** books, including:

- Basic astronomy concepts
- Basic chemistry concepts
- Balancing chemical reactions
- Calculus-based physics textbooks
- Calculus-based physics workbooks
- Calculus-based physics examples
- Trig-based physics workbooks
- Trig-based physics examples
- Creative physics problems

www.monkeyphysicsblog.wordpress.com

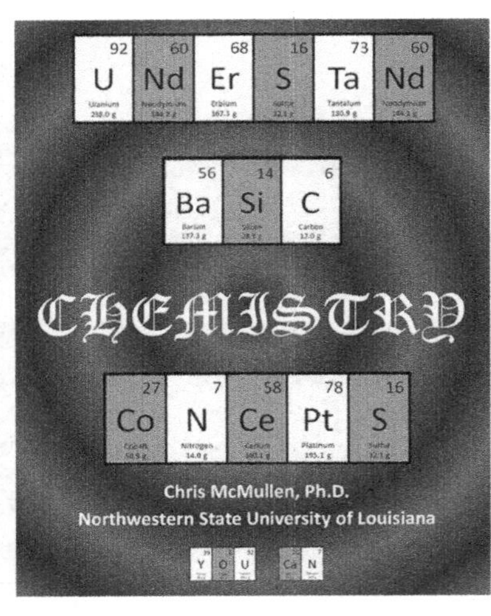

THE FOURTH DIMENSION

Are you curious about a possible fourth dimension of space?

- Explore the world of hypercubes and hyperspheres.
- Imagine living in a two-dimensional world.
- Try to understand the fourth dimension by analogy.
- Several illustrations help to try to visualize a fourth dimension of space.
- Investigate hypercube patterns.
- What would it be like to be a four-dimensional being living in a four-dimensional world?
- Learn about the physics of a possible four-dimensional universe.

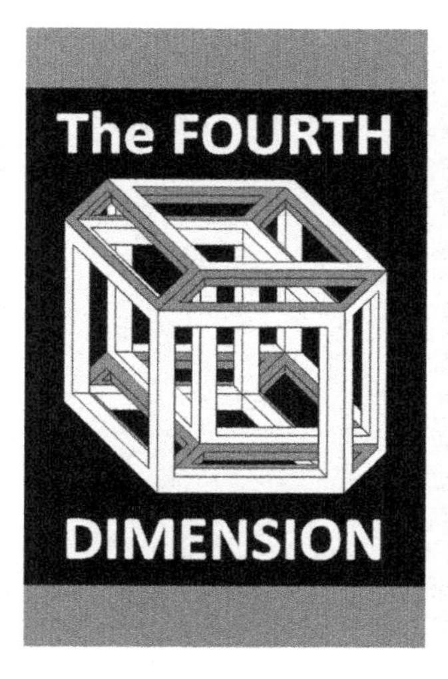

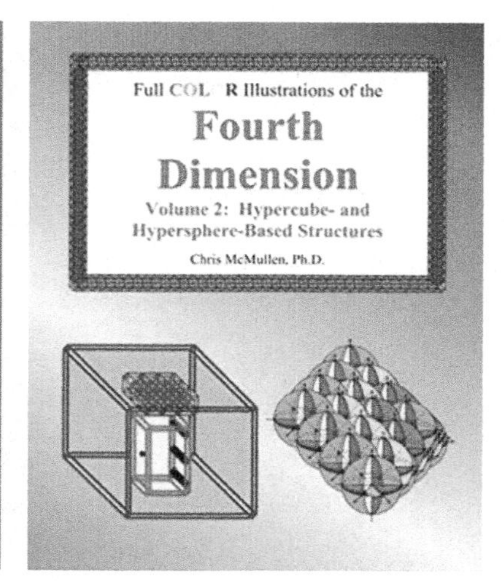

Made in the USA
Middletown, DE
04 March 2021